T0318869

PEACOCK BASS

PEACOCK BASS

Diversity, Ecology and Conservation

Kirk O. Winemiller
Department of Ecology and Conservation Biology, Texas A&M University, College Station, TX, United States

Leslie C. Kelso Winemiller
Department of Ecology and Conservation Biology, Texas A&M University, College Station, TX, United States

Carmen G. Montaña
Department of Biology, Stephen F. Austin State University, Nacogdoches, TX, United States

Academic Press is an imprint of Elsevier
125 London Wall, London EC2Y 5AS, United Kingdom
525 B Street, Suite 1650, San Diego, CA 92101, United States
50 Hampshire Street, 5th Floor, Cambridge, MA 02139, United States
The Boulevard, Langford Lane, Kidlington, Oxford OX5 1GB, United Kingdom

Cover image courtesy of Kirk O. Winemiller. The paintings that appear in this book are original Guyotaku artwork by Kirk O. Winemiller. Guyotaku is the ancient Japanese method of blotting a fish with ink and then pressing it with rice paper to capture the fish image, including features of scales, spines and fin rays. Ink rubbings are then photocopied onto paper suitable for painting with water colors. The same fish image was used for paintings of the nine peacock bass species and the cover of this book.

Library of Congress Cataloging-in-Publication Data
A catalog record for this book is available from the Library of Congress

British Library Cataloguing-in-Publication Data
A catalogue record for this book is available from the British Library

ISBN: 978-0-323-85157-2

For information on all Academic Press publications
visit our website at https://www.elsevier.com/books-and-journals

Publisher: Megan R. Ball
Acquisitions Editor: Patricia Osborn
Editorial Project Manager: Allison Hill
Production Project Manager: Selvaraj Raviraj
Cover Designer: Mark Rogers

Typeset by SPi Global, India

Contents

Preface vii

1. The alluring peacock bass
Natural history 4
Competitors, predators, and parasites 15
Traits useful for identification of peacock bass species 23
References 31

2. Butterfly peacock bass, *Cichla ocellaris* (Bloch & Schneider 1801)
Identification 39
Distribution and habitat 46
Population structure and abundance 52
Feeding 53
Growth 56
Reproduction 57
References 62

3. Orinoco butterfly peacock bass, *Cichla orinocensis* (Humboldt, in Humboldt & Valenciennes, 1821)
Identification 68
Distribution and habitat 72
Feeding 78
Growth 82
Population abundance and structure 84
Reproduction 84
References 88

4. Royal peacock bass, *Cichla intermedia* (Machado-Allison, 1971)
Identification 94
Distribution and habitat 97
Feeding 99
Growth 100
Population abundance and structure 101
Reproduction 102
References 103

5. Speckled peacock bass, *Cichla temensis* (Humboldt, in Humboldt & Valenciennes, 1821)
Identification 111
Distribution and habitat 115
Feeding 121
Growth 125
Population abundance and structure 126
Reproduction 130
References 134

6. Pinima peacock bass, *Cichla pinima* (Kullander & Ferreira 2006)
Identification 139
Distribution and habitat 143
Feeding 149
Growth 149
Population abundance and structure 150
Reproduction 150
References 152

7. Blue peacock bass, *Cichla piquiti* (Kullander & Ferreira 2006)
Identification 157
Distribution and habitat 161
Feeding 165
Growth 166
Population abundance and structure 167
Reproduction 169
References 170

8. Xingu peacock bass, *Cichla melaniae* (Kullander & Ferreira 2006)
Identification 178
Distribution and habitat 180
Feeding 184
Growth 185
Population abundance and structure 187
Reproduction 188
Impacts from hydropower 189
References 192

9. Fire peacock bass, *Cichla mirianae* (Kullander & Ferreira 2006)

Identification 199
Distribution and habitat 201
Feeding 205
Growth 210
Population abundance and structure 210
Reproduction 212
Reference 214

10. Falls lukunani, *Cichla cataractae* (Sabaj, López-Fernández, Willis, Hemraj, Taphorn & Winemiller 2020)

Identification 218
Distribution and habitat 222
Feeding 225
Growth 225
Population abundance and structure 226
Reproduction 227
References 229

11. Evolutionary relationships and zoogeography

Evolutionary relationships 234
Zoogeography 240
Hybridization 263
References 266

12. Fisheries, captive care, and conservation

Fisheries—Native stocks 272
Fisheries—Reservoirs 277
Introduced stocks 279
Care of captive peacock bass 282
Fisheries conservation 287
Epilogue 296
References 297

Index 301

Preface

With all the interesting fishes in the world, why write a book about peacock bass? Quite simply, peacock bass are our favorite fish among the hundreds of species that the three of us have studied during a collective century of field research on rivers and lakes worldwide. What makes this group of fishes so appealing? For most people the answer would be the peacock bass's reputation as one of the world's premier sport fish, and we certainly would agree with this viewpoint. Among freshwater and marine fishes pursued by anglers, the peacock bass is renowned for the ferocity of its attack, aerial acrobatics, brute strength, and wily escape tactics that all too frequently frustrate even the most accomplished anglers. However, peacock bass are more, considerably more—much more—than just an awesome sport fish. Like other species of the Cichlidae, a large and diverse family of tropical freshwater fishes, peacock bass (species of the genus *Cichla*) have fascinating reproductive biology that includes aggressive defense of eggs and fry by both parents. Many an angler's lure has been violently attacked as it slices through the water in close proximity of a brooding pair. Yet the best way to appreciate the parenting skills of peacock bass is to observe them up close underwater. To witness the boldness and diligence of a pair of peacock bass tending their brood is to experience one of nature's true wonders.

With vibrant colors and elaborate markings that rival those of coral reef fishes, peacock bass are among the most beautiful of all freshwater fishes. Peacock bass color patterns show impressive variation both among and within individual species, and this variation has greatly confused biologists, naturalists, anglers, and even local fishermen. Indeed, this confusion was one of the major motivations for writing this book. Color patterns vary from species to species in ways that are sometimes obvious but oftentimes quite subtle. New research using genetic data is revealing species boundaries, and this book attempts to summarize the state of *Cichla* taxonomy so that color patterns can be properly interpreted. Color patterns also vary within local populations, sometimes strikingly, in accordance with the fish's age and reproductive status or water chemistry and seasonal variation in water level. Species of the genus *Cichla* also are providing scientists with an excellent model system for studying population genetics and evolution. New studies are revealing the influence of biogeography, species hybridization, physiology, behavior, and ecology on the evolutionary diversification of the lineage.

There is little doubt that peacock bass species are some of the most ecologically and economically important fishes in South America where they are native. Given their popularity with anglers, peacock bass have been introduced to waters outside their native range, including lakes and canals in North and Central America, Asia, Hawaii, and Puerto Rico. Their ecological importance stems from their widespread distribution in lowland rivers of South America, generally high abundance, and voracious predatory nature. Research has revealed that peacock bass in rivers and lakes can exert strong control over prey-fish stocks and thereby have the potential to function as keystone predators, i.e., species that play critical roles in ecosystem dynamics. As a keystone species, peacock bass likely enhance biodiversity in their native habitats, but when people introduce peacock bass into waters outside their natural range, they can have a negative impact on fish diversity and ecological processes. Peacock bass are considered one of the most delicious freshwater fishes to eat and have high value in fish markets, which oftentimes results in overfishing. Management of peacock bass fisheries is highly variable, and enforcement of regulations varies greatly from well-controlled, community-based fisheries to virtual lack of enforcement over vast areas. Ecotourism for peacock bass fishing has emerged as a significant economic component in South America, and its immense popularity is evident from the myriad websites displaying accounts, photos, and videos.

This book summarizes all that is currently known about our favorite fishes, the peacock bass, a voracious yet beautiful tropical predator. These fascinating fish rank among the most scientifically, ecologically, and economically valuable fishes inhabiting tropical freshwaters. This book describes the diversity and ecology of peacock bass plus brief accounts for a variety of other fish and aquatic animals found in rivers and lakes of South America. Our information and insights are based on our scientific field experiences plus the growing scientific literature on *Cichla* species. This book provides only a few tips about how to catch peacock bass and makes no attempt to reveal the best fishing locations; other books are available to guide anglers in those areas. We hope, however, that this book will be of great interest to anglers who wish to gain a better understanding of, and appreciation for these amazing fishes.

We owe a great deal of gratitude to professional colleagues who assisted us with research and contributed logistics, funding, information, and enjoyable times catching fish in remote regions of South America. Among the friends, family, colleagues, and others who contributed to this book, we acknowledge Stuart Willis, Donald Taphorn, Aniello Barbarino Duque, Basil Stergios, Octaviano Santaella, the late Oscar Leon Mata, the late George and Carolyn Wierichs Kelso, Brent Winemiller, Megan Winemiller, Edgar Pelaez, Carol Marzuola, Glen Webb, Clayton Lofgren, the late Tom Nash, Amy Nash, Dave Jepsen, Tammy McGuire, Albrey Arrington, Craig Layman, José Garcia, Domingo Garcia, Nola Blanco, David Hoeinghaus, Hernán López-Fernández, Karen Alofs, José Vicente Montoya, Dan Roelke, Chris Peterson, Eric Pianka, Lee Fitzgerald, Douglas Rodriguez, Mark Sabaj, John Lundberg, Nathan Lujan, Dan Fitzgerald, Caroline Arantes, Katie Roach, Bibiana Correa, Carlos Lasso Alcala, Crispulo Marrero, Frank Ibarra, Jay Roff, Alexis Medina, Angelo Agostinho, Luiz Carlos Gomes, João Latini, Fernando Pelicice, Luciano Montag, Marcelo Andrade, Jansen Zuanon, Moacir Fortes Senior (Mo Sr.), Moacir Fortes Junior (Mo Jr.), Paulo Petry, Pedro Gull, Ashley Holland, Devin Bloom, Guillermo Ortí, Ryan King, Michi Tobler, the late Phil Cochran, Jenny Cochran Biederman, Alex Flecker, Tom Turner, Frank Pezold, Leo Nico, Howard Jelks, Steve Walsh, Erling Holm, Brad Pusey, Allison Pease, Clint Robertson, Kevin Mayes, Anne Ashmun, John Karges, Kevin Conway, Eduardo Cunha, José Birindelli, Renato Silvano, Daniel González, Leonardo Okada, Cristian Venegas, Wesley Wong, and students from the Amazon River Tropical Biology study abroad course at Texas A&M University.

CHAPTER

1

The alluring peacock bass

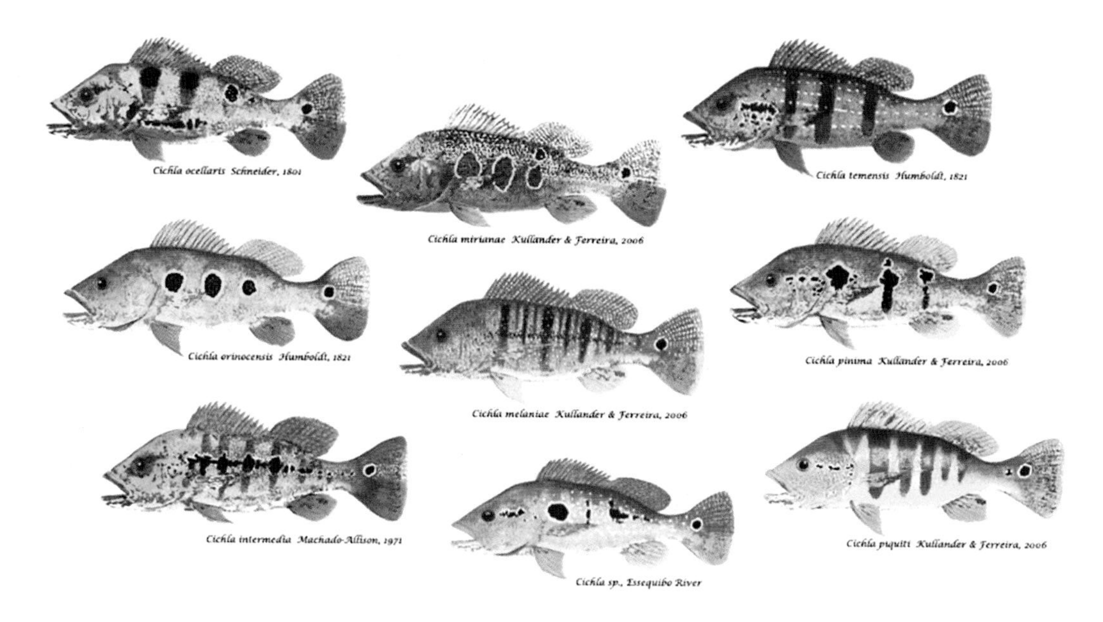

With the tropical sun intensely radiating on my back, I (KW) was sitting in a small aluminum boat next to a sandy bank of the Río Cinaruco in Venezuela's Santos Luzardo National Park. As I repeatedly cast a lure near a fallen brush pile, I chatted with my graduate students Dave Jepsen and Tammy McGuire about their research projects. Recently married, Dave and Tammy had been living in the national park and collecting data for their research projects—his on the ecology of three species of pavón, as the Venezuelans call peacock bass, and hers on the ecology of pink river dolphins (*Inia geoffrensis*). Peacock bass and river dolphins are both abundant in the Río Cinaruco, but on this morning, we were getting few bites from the former and seeing only a few individuals of the latter swimming near our boat.

All of a sudden, the water surface in the middle of the river channel began to boil and swirl as a group of large predators attacked a school of fish. Ensuing not more than 100 m from the boat was a succession of violent explosions—pop, pop, pop! I turned to Tammy and naively asked, "dolphins?" She replied, "No, the dolphins never feed like that." We quickly realized that the attackers were a school of hungry peacock bass, most likely "pavón cinchado," the

Peacock Bass
https://doi.org/10.1016/B978-0-323-85157-2.00010-X

largest of the three peacock bass species inhabiting the river. I shouted to Dave, our boat operator, "Go, go, go!" The first lure to hit the water was immediately engulfed by a hungry peacock bass, as was each lure on successive casts. Because Dave needed fish measurements for his research, he had to pause from fishing to collect the precious data before releasing each fish back into the river, but I kept pulling in peacock bass until the action halted almost as abruptly as it had started. We had just witnessed a peacock bass feeding frenzy, one that was no less violent than those performed by bluefish, tuna, and other voracious marine predators. After catching seven impressive "pavón cinchados" (or banded peacock bass, *C. temensis*—a good guess as it turned out), we motored our boat along the shoreline where we noticed hundreds of fish migrating upstream in a continuous, shadowy column. These fish were young-of-the-year bocachicos (*Semaprochilodus kneri*), silvery fish with yellow and black striped tails that had migrated into the Cinaruco River from the Orinoco where they had hatched and spent the wet season feeding within productive aquatic habitats in the vast floodplains. During the early stages of the annual dry season, thousands of young bocachicos migrate and eventually take up residence in the Cinaruco River and other clearwater tributaries of the Orinoco. Dave's research would later confirm that peacock bass take full advantage

Aerial photograph of the Cinaruco River in the Llanos region of Venezuela. The Orinoco River can be seen at the top of the photo. The entire landscape in this photo floods each year during the wet season. *Photo: K. Winemiller.*

The Cinaruco River is a clearwater tributary of the Orinoco that supports three peacock bass species. *Photo: K. Winemiller.*

A fine speckled peacock bass (*C. temensis*) caught on a surface lure fished over a sandbar of the Cinaruco River in the Llanos region of Venezuela. *Photo: K. Winemiller.*

of this bountiful food supply. Based on analysis of gut contents as well as carbon and nitrogen stable isotope ratios of fish tissues, he estimated that, on average, bocachicos comprise about 50% of the diet of resident large *C. temensis* during the migration period.

This story reveals two important points about peacock bass. First, they rank among the most voracious predators inhabiting freshwaters anywhere on earth and thus illicit a significant adrenalin rush for anyone inclined to entice them with a lure and then hang on for the ride. Second, there is much we don't know about these amazing fish or the ecology of the tropical rivers where they live, and therefore, any amount of time spent researching them is certain to reward the investigator with a few surprises. Dave and I ultimately determined that migrating bocachicos provide peacock bass with what's called a *spatial food-web subsidy*. This nutritional windfall occurs when food resources are transferred from a productive ecosystem, such as the Orinoco floodplains, into an unproductive ecosystem like the Cinaruco River. This imported food is consumed by resident predators of the unproductive system, thereby allowing their populations to increase to levels otherwise unattainable. If sufficiently large, spatial food-web subsidies can result in a *trophic cascade*, whereby predator populations buoyed by an influx of food resources originating from outside the system begin to exert stronger control on resident prey populations.

A 1-year old bocachico (*Semaprochilodus kneri*), a migratory fish that is common in the Cinaruco and other clearwater rivers of the Orinoco Basin during the dry season. Young bocachicos are an important prey of peacock bass. *Photo: K. Winemiller.*

Natural history

A wonderful aspect about studying and fishing for peacock bass is the places where you must travel to find them. Peacock bass are native to rivers and lakes of the Amazon, Orinoco, and drainage basins along the northeastern coast of South America. Peacock bass inhabit lowland habitats, especially meandering rivers with broad floodplains subject to inundation during the wet season. A few species inhabit rocky shoals with swiftly flowing water in rivers of the Guyana and Brazilian shield regions (ancient geologic formations in northern and northeastern portions of the continent). Thus anglers in pursuit of peacock bass in their native habitat also enjoy the experience of exploring jungles and savannas of the Neotropics that are home to some of the planet's most impressive biological diversity. A journey in search of peacock bass is likely to yield sightings of monkeys, caiman, giant river otters, anacondas, river dolphins, capybara, and innumerable species of birds, insects, and plants. For us, nothing quite compares with the experience of floating over tea-stained waters swirling past boulders and logs with banks cloaked by lush tropical forest and surrounded by the calls of howler monkeys, macaws, and ibis. The native range of the peacock bass also happens to be where the world's greatest diversity of freshwater fishes is found, which of course is a powerful attraction for curious ichthyologists or aquatic ecologists.

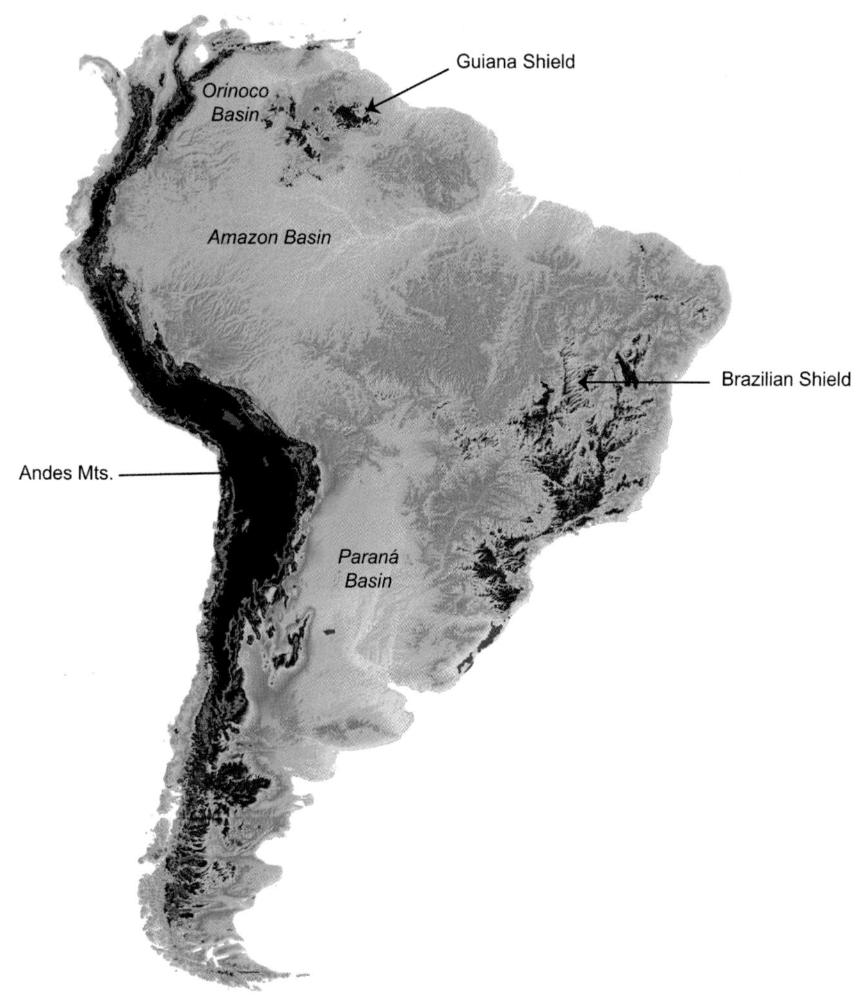

Digital elevation map showing locations of two ancient highlands, the Guiana and Brazilian shields, and the much younger Andes Mountains. Clearwater and blackwater rivers dominate the shield regions, and turbid whitewater rivers drain the Andes.

Major river basins of South America where peacock bass are found. Peacock bass are not native to the Magdalena, Maracaibo, São Francisco, and Paraguay-Paraná basins. Four species co-occur naturally within the Casiquiare River drainage, a river that links the Orinoco and Negro rivers.

Peacock bass have become increasingly popular among aquarists dedicated to the care and breeding of large, impressive fish. In recent years, postings of photos and videos of peacock bass housed in aquaria have proliferated on the internet. One can find aquarists' video clips for most of the peacock bass species, including sequences showing spawning and brood care. The use of small digital cameras that can be submerged underwater has resulted in a proliferation of internet postings of photos and videos of peacock bass in their natural surroundings. There even is a new kind of ecotourism that has emerged among aquarists—international excursions that cater to aquarium hobbyists wanting to see tropical fishes in their natural habitats. Habitats where peacock bass are found often contain hundreds of other fish species, many of them familiar to aquarists. These include small ornamental fishes, such as tetras (Characidae), headstanders (Anostomidae), and plecos (armored catfishes belonging to the family Loricariidae) as well as other kinds of cichlids, such as pike cichlids (*Crenicichla* species),

A lake in the floodplain of the Pasiba River, a tributary of the Casiquiare River in Venezuela's Amazonas state. Four species of peacock bass can be found in this river. *Photo: C. Montaña.*

The black piranha (*Serrasalmus rhombeus*) frequently is found in habitats with peacock bass, where these two predators likely feed on each other's juveniles but probably interact only infrequently as adults. *Photo: K. Winemiller.*

flag cichlids (*Mesonauta* species), eartheaters (*Geophagus* and *Satanoperca* species), and dwarf cichlids (*Apistogramma* species). Rivers and floodplain lakes that support peacock bass also harbor other kinds of large predatory fishes, such as piranhas (Serrasalmidae), wolf fish (Erythrinidae), vampire fish (Cynodontidae), and large catfishes (Pimelodidae).

A red pike cichlid from the Pasimoni River, Venezuela, either a variety of *Crenicichla lugubris* or an as-yet-undescribed species. *Photo: K. Winemiller.*

For this book, we elected to use the name "peacock bass" for fishes belonging to the genus *Cichla*. Peacock bass have various common names that depend on the species, region, language, and sometimes even the state of maturation and color pattern of specimens belonging to the same species. In Brazil they are called "tucunaré," an indigenous term meaning "handsome fish." In Venezuela and Colombia, *Cichla* are called "pavón," Spanish for "peacock." This name derives from the distinct ocellus (a black spot bordered by a bright yellow or white ring that resembles an eye spot) that is always present near the base of the caudal fin (tail fin) of all species. This caudal ocellus resembles the ocelli present on the tail-feathers of the male peacock. Consequently, most English speakers have adopted the name "peacock bass," which seems to be the most commonly used term among international anglers, aquarium hobbyists, and naturalists. Of course, these fish are not actually a type of "bass" (a name for fish from several freshwater and marine families), rather they are cichlids (i.e., members of the family Cichlidae). In Guyana, where English is the official language, the indigenous name "lukunani" is used for peacock bass. Names used for *Cichla* by various indigenous groups of French Guiana include "aboné," "tukunali," "malisamba," "kunan," "matawalé," and "toekoenari." In addition to widespread use of the name "tucunaré" in Brazil, people in different regions of the Amazon use their own local names for *Cichla*, including "furiba," "peixe modeda," and "pitanga." In Panama, the *Cichla ocellaris* introduced into Lake Gatún are called "sargento," and *Cichla* introduced into reservoirs of Puerto Rico are called "mamito" by the locals. When more than one species occurs in a region, the local people use adjectives to distinguish the various forms, usually based on coloration patterns. For example, in Venezuela we find the "pavón mariposa" (butterfly peacock bass, *C. orinocensis*), "pavón royal" (royal peacock bass, *C. intermedia*), "pavón lapa" (speckled peacock bass, *C. temensis* in immature or adult nonreproductive state), and "pavón cinchado" (banded peacock bass, *C. temensis* with banded pattern of the adult in reproductive state).

In the chapters that follow, we summarize currently available information on the diversity, geographic distributions, ecology, and conservation of the various species of peacock bass. There remains disagreement among ichthyologists regarding the number of valid species in the genus *Cichla*. A major taxonomic revision of the genus performed by Sven Kullander and Efrem Ferreira [1] included descriptions of nine new species from the Amazon Basin, which brought the total number of peacock bass species to 15. (Hereafter, whenever we reference this important taxonomic revision by Kullander and Ferreira (2006), we use the abbreviation "K&F.") Recent research of *Cichla* population genetics and phylogenetics done by Stuart Willis and colleagues [2–4] recognized only eight valid species. These scientists concluded that some of the new species described by K&F and at least two previously described species actually represent regional populations of the same species that have different coloration patterns and morphology. Geographic distributions and evolutionary relationships of *Cichla* species and regional varieties will be discussed in more detail in the chapters that follow. Following the findings of Willis and collaborators, we recognize only eight valid species among the 15 that were proposed by K&F, with two of them, *Cichla ocellaris* and *C. pinima*, having regional populations that may be designated as *varieties* (or *subspecies*). These varieties are broadly overlapping in their coloration and morphology and seem to have genetic patterns that are not sufficiently distinct to warrant recognition as separate species. Therefore, we designate four regional varieties of the species *Cichla ocellaris* (*C. ocellaris* var. *ocellaris*, *C. ocellaris* var. *kelberi*, *C. ocellaris* var. *monoculus* [which includes fish referred to as *C. nigrolineata*], and *C. ocellaris* var. *pleiozona*). Four of the new species described by Kullander and Ferreira (*C. pinima*, *C. jariina*, *C. thyrorus*, and *C. vazzoleri*) do not appear to correspond to genetically distinct lineages, and therefore are grouped together as a single species *Cichla pinima*, the most widely distributed of the four varieties. The previously described species recognized by Willis and collaborators as valid based on genetics analysis are *Cichla intermedia*, *C. orinocensis*, and *C. temensis*. Among the species newly described by K&F, those recognized as valid are *Cichla melaniae*, *C. mirianae*, and *C. piquiti*. Finally, we recognize a ninth species, *Cichla cataractae*, the falls lukunani from the Essequibo River Basin of Guyana, that was only recently described (i.e., the species name was officially designated within a study published in a recognized scientific journal). In Chapter 11, we further discuss the potential for discovery of more *Cichla* species as well as evidence of hybridization between species.

A fire peacock bass from the Sáo Benedito River, Brazil. *Photo: K. Winemiller.*

The various species of peacock bass generate intense interest not only from biologists, aquarists, and anglers, but also subsistence and commercial fishers, fisheries managers, and conservationists. The final chapter of the book summarizes information about fisheries and management of peacock bass stocks, both in regions where they are native and those where they have been introduced. As apex (top) predators, peacock bass are sensitive indicators of the overall status of the ecosystem. Environmental impacts that affect species at lower positions in the food web, such as habitat degradation or pollution, ultimately will be reflected by changes in the abundance or stock structure of apex predators. Therefore peacock bass are not only beautiful for their colors, interesting for their behavior, important for their ecology, and prized for food and recreation, but also are useful as sentinels of environmental impacts caused by human actions.

Despite the fact that peacock bass are among the most common and widespread predatory fish in South America, knowledge about their ecology has been limited until recently. Most of the early information about feeding, growth, and reproduction in natural populations emerged from research conducted in Venezuela on three species—*Cichla intermedia*, *C. orinocensis*, and *C. temensis* [5–8]. Ecological information from rivers in other regions within the native range of peacock bass was scarce until fairly recently [9, 10]. Research also has been conducted on peacock bass stocked in aquaculture ponds [11] and reservoirs in South America [12–15], Panama [16, 17], Puerto Rico [18], and Hawaii [19] plus canals of south Florida [20].

Peacock bass inhabit diverse habitats, ranging from creeks to large rivers, natural floodplain lakes, and constructed reservoirs. Although the various peacock bass species have fairly distinct habitat affinities, general requirements appear to be high water transparency, warm temperatures, and access to nesting habitats with still or slowly flowing water [6]. Peacock bass are pursuit predators that use vision as their principal means to locate and track prey, and this explains their absence from turbid waters with low visibility.

In their native range, peacock bass tend to be most common in clearwater and blackwater creeks, rivers, and lakes. It was the 19th century naturalist, Alfred Russel Wallace who first recognized the great variation in the water properties of tropical rivers and its effects on biota. He first coined the terms *clearwater*, *blackwater*, and *whitewater*. In the Amazon, *clear waters* originate from the ancient weathered geology of the Brazilian Shield and other areas with relatively poor soils lacking extensive wetlands [21]. These waters tend to be highly transparent, sometimes with green tones when phytoplankton are abundant, often with slightly acidic pH. In contrast, *black waters* have high concentrations of humic and fulvic acids derived from decaying vegetation, which gives them a tea-stained appearance and very low pH. Black waters generally flow from poorly drained areas with deep sandy soils overlying clay or other impermeable substrates and normally have extremely low concentrations of inorganic ions and nutrients as well as suspended particles that cause turbidity [21, 22]. Even though black waters lack turbidity, concentration of dissolved humic substances is sometimes so high that the water appears dark red when viewed up close and black when viewed from a distance. For a fish or human diver, it is difficult to see any object located more than a half a meter away. This fact is important for anglers to consider when selecting lures for peacock bass, and the best option in extreme black water is to use lures that are opaque white or yellow, nonreflective, and capable of splashing at the surface.

White waters of the Neotropics are turbid with high loads of suspended particles of clay and other materials, and they tend to have neutral pH. In the tropics, white waters are not necessarily rapidly flowing, which often is a source of confusion given the popular use of the term for describing rapids and raging rivers. In the Amazon and Orinoco basins, most whitewater rivers originate in the Andes Mountains, a relatively young land formation that is being actively weathered to this day. Compared to clearwater and blackwater rivers in South America, these muddy rivers have relatively high concentrations of ions and nutrients and very low light transmission that makes them unsuitable for peacock bass. However, some species of peacock bass do well in lagoons and creeks in the floodplains of whitewater rivers, including the mainstem Amazon (called the Solimões in Brazil in the reach upstream from the confluence of the Rio Negro), where suspended sediments settle from the water column and result in greater transparency. Table 1.1 summarizes differences in water chemistry of some rivers and reservoirs that support peacock bass.

White sand bank emergent from dark black water of the Rio Negro. *Photo: K. Winemiller.*

The clearwater Aguaro River in the Venezuelan Llanos provides suitable habitat for peacock bass. *Photo: K. Winemiller.*

The Apure River, a whitewater river in the Venezuelan Llanos, carries heavy loads of suspended clay, silt, and other sediments that wash down from the Andes Mountains. The turbidity of whitewater rivers makes them unsuitable for peacock bass. *Photo: K. Winemiller.*

Muddy water of the Amazon River is unsuitable for peacock bass and probably hinders dispersal among clearwater tributaries. *Photo: K. Winemiller.*

TABLE 1.1 Water quality parameters of tributaries of the Orinoco, Amazon, and Essequibo river basins (top panel). Water quality parameters of reservoirs where peacock bass have been introduced (bottom panel). Values are based on averages measured during lower-water periods in the river channel.

River channel	pH	Conductivity (μmhos/cm)	Transparency (m)	Dissolved oxygen (mg/L)	Water type
Rio Japura—Brazil	6.9	67	0.3	–	White
Solimões—Brazil	6.7	92	0.4	–	White
Rio Negro—Brazil	4.7	11	0.5	4.9	Black
Pasimoni—Venezuela	4.6	30	0.5	2.4	Black
Casiquiare—Venezuela	5.5	39	1.2	3.8	Black-clear
Siapa—Venezuela	5.4	21	1.4	–	Black-clear
Pasiba—Venezuela	5.5	23	–	–	Black
Ventuari—Venezuela	5.5	10	1.6	6.6	Clear
Upper Orinoco—Venezuela	6.1	10	0.7	–	White
Cinaruco—Venezuela	5.7	11	1.3	6.4	Black-clear
Apure—Venezuela	7.7	200	0.2	6.3	White
Lower Orinoco—Venezuela	7.1	60	0.3	6.1	White
Caura—Venezuela	5.6	13	0.7	6.7	Black
Essequibo—Guyana	6.1	–	1.2	–	Black-clear
Paragua—Bolivia	5.3	–	1.3	–	Clear
Madre de Dios—Peru	7.3	116	<0.2	6.6	White

Reservoirs	pH	Conductivity (μmhos/cm)	Transparency (m)	Dissolved oxygen (mg/L)	Water type
Guri, Venezuela	6	27	2.3	6	Black
La Coromoto, Venezuela	6.6	139	3.1	6.3	Clear
Masparro, Venezuela	7.9	105	3.9	8.9	Clear
Três Marias, Minas Gerais, Brazil	7.8	55	3.7	6.5	Clear
Cachoeria Dourada, Paranaiba River, Brazil	5.9	30	–	–	Clear
Ribeirão das Lajes, Rio de Janeiro, Brazil	7.1	29	2.9	–	Clear
Lake Gatun, Panama	7.0	73	–	4.5	Clear

Competitors, predators, and parasites

If fish were track athletes, peacock bass would be regarded as champion sprinters. Peacock bass attack prey with an awe-inspiring combination of speed and power. Many an angler has watched in amazement as his or her lure was obliterated by a ravenous peacock bass. "Explosion" is a word used frequently to describe peacock bass attacks on lures. It's equally impressive to observe peacock bass feeding underwater. We have fed live fish to aquarium-housed peacock bass as well as numerous other kinds of predatory fish, including largemouth bass, smallmouth bass, tarpon, piranhas, payaras, and shovelnose catfish. The incredible speed, accuracy, and downright viciousness of the peacock bass attack are unparalleled. Most times, the fleeing prey disappears in the blink of any eye as the peacock bass bolts toward the surface. Because they are elite sprinters, these bursts of power use a lot of energy. After one or two attacks, the fish's ventilation rate increases for a while (as observed by the opening and closing of the mouth and gill covers to draw oxygenated water over the gills) to repay their oxygen debt. Sprinters have power, but they lack endurance; maximum speed and power can't be sustained for long periods. Consequently, peacock bass don't have the endurance to make long runs and sustain fights with anglers for an hour or more in the manner of a tarpon, tuna, or billfish, for example. However, when angling for peacock bass, it's important to keep a tight grip on the fishing rod. When a strike comes to the unprepared angler, either the rod, the person, or both might end up being yanked into the water.

There are other large predators that coexist with peacock bass in the rivers and lakes of South America, and of course other parts of the world have their own large predatory fishes. In the Amazon, Orinoco, and Essequibo rivers, peacock bass share the waters with giant catfishes (*Brachyplatystoma filamentosum*, *Phractocephalus hemilioipterus*, *Zungaro zungaro*), pirarucú (*Arapaima gigas*), and toothy vampire fish (*Hydrolycus armatus*, *H. scomberoides*, also called dogtooth characin or payara) and wolf fish

The payara (*Hydrolycus armatus*) has fangs on the lower jaw that it uses to impale its prey. This one was caught in Venezuela's Cinaruco River. *Photo: K. Winemiller.*

(*Hoplias aimara*, *H. malabaricus*, also called aimara or taira). In Africa, the largest predatory fishes are tigerfish (*Hydrocynus* spp.), aba (*Gymnarchus niloticus*), and vundu catfish (*Heterobranchus longifilis*), and Asia has its own river monsters, including the goonch catfish (*Bagarius yarrelli*), giant pangasius catfish (*Pangasius sanitwongsei*), and giant snakehead (*Channa micropeltes*). Large predatory fishes also are found in temperate freshwaters of North America (e.g., gar, pike, catfish, and bass) and Europe (e.g., wels catfish, *Silurus glanis*). However, the Amazon and Orinoco river basins are home to the most diverse freshwater fish faunas on Earth, and this includes predatory fishes large and small. Some of these predators are highly specialized and feed during either daytime or nighttime and only on certain kinds of fish of a particular size. Peacock bass are active during daytime as well as dietary generalists that feed on nearly any kind of fish that will fit into their large mouths.

The aimara (*Hoplias aimara*) has sharp teeth and powerful jaws that close like a bear trap. This fish was caught in Guyana's Rewa River. *Photo: K. Winemiller.*

The pirarucú, or arapaima (*Arapaima gigas*), is an important food fish in Brazil, and recently has become highly sought after by anglers wishing to catch a "river monster." *Photo: K. Winemiller.*

Peacock bass coexist with other kinds of aquatic top predators besides fish, including the aforementioned pink river dolphin (*Inia geoffrensis*) referred to as *boto* in Brazil and *tonina* in Venezuela and Colombia. Other aquatic predators include the smaller gray river dolphin called tucuxi (*Sotalia fluviatilis*), giant river otter (*Pteronura brasiliensis*), green anaconda (*Eunectes murinus*), and various species of caiman (*Caiman crocodilus*, *Melanosuchus niger*, *Paleosuchus palpebrosus*). Also competing with peacock bass are numerous fish-eating birds, including ospreys, anhingas, herons, storks, and kingfishers.

Green anaconda (*Eunectes murinus*) found basking on the shore of the Rio Branco, Brazil. *Photo: K. Winemiller.*

The giant river otter may be the only aquatic predator that can match the peacock bass in its voracity and skill in capturing fish. This individual was swimming with its family in the Rio Branco, Brazil. *Photo: K. Winemiller.*

Common in rivers and floodplain lakes throughout the Amazon and Orinoco basins, pink river dolphins (*Inia geoffrensis*), like these in the Rio Negro, are proficient at catching and consuming a great variety of fish, including peacock bass. *Photo: K. Winemiller.*

Along with peacock bass, giant river otters, and pink river dolphins, black caiman (*Melanosuchus niger*), such as this one from the São Benedito River in Brazil, are apex predators in rivers and lakes of the Amazon. These crocodilians may grow to 6 m (20 ft). *Photo: K. Winemiller.*

The Orinoco crocodile (*Crocodylus intermedius*) from the Cinaruco River is restricted to the Llanos region of the Orinoco Basin and is one of the world's largest crocodilians. Due to overhunting for its valuable skin, this apex predator is now one of the world's most endangered reptiles. *Photo: K. Winemiller.*

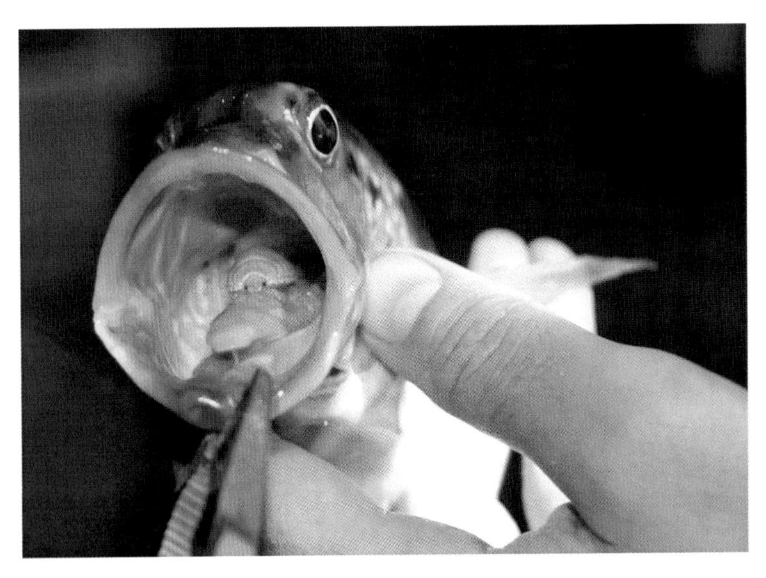

Peacock bass from the lower Xingu River with a parasitic isopod in its mouth. *Photo: N. Lujan.*

Peacock bass in the wild generally have parasites, especially nematode worms that inhabit their gut. These worms don't seem to cause much harm, and sometimes dozens of these parasites are found in a fish's gut. Other kinds of nematodes sometimes infect the eye, liver, gonads, and other organs, and these can significantly impact the fish's health. Bacterial and fungal infections are sometimes observed on the body and fins. Perhaps the strangest parasite of peacock bass is the tongue-eating louse, an isopod in the family Cymothoidae. These crustaceans attach to the fish externally and then enter the gill opening before finding their way into the fish's mouth. Once inside the mouth, it penetrates the tongue with its appendages, causing the tongue to atrophy. The isopod has the same shape as the tongue and apparently replaces its mechanical function during prey ingestion. The parasite feeds on the host's blood and mucus but apparently does minimal harm.

Various types of bacteria, fungi, protozoa, worms, and other types of invertebrates parasitize fish, but there is one type of vertebrate parasite, the tiny candirú catfish. The little bloodsuckers are members of the South American catfish family Trichomycteridae, sometimes referred to as pencil catfish. Within this species-rich family, several genera, including *Stegophilus*, *Tridensimilus*, and *Vandellia* are bloodsuckers that enter the gill chamber of fish where they attach to the gill arches and filaments using tiny hooks on their heads. There they chew into the fish's blood-rich gill filaments to obtain nourishment. A variety of fish are attacked by candirú, among them peacock bass and other cichlids. The candirú is infamous not only for its grotesque blood-feeding habit, but also for being the subject of hideous legends about tiny fish entering and wiggling their way up the urethras of unfortunate people who urinated while bathing in South American rivers. Except these are not legends, and there are documented cases of individuals receiving medical attention for infection and pain caused by a tiny catfish lodged in their urethra [23]. Another species of tiny trichomycterid catfish is a parasite that feeds on the protective layer of mucus slime that coats the bodies of most fish [24]. During daytime, these tiny fish live buried in the sand, and they emerge at night to graze on fish sleeping in shallow water.

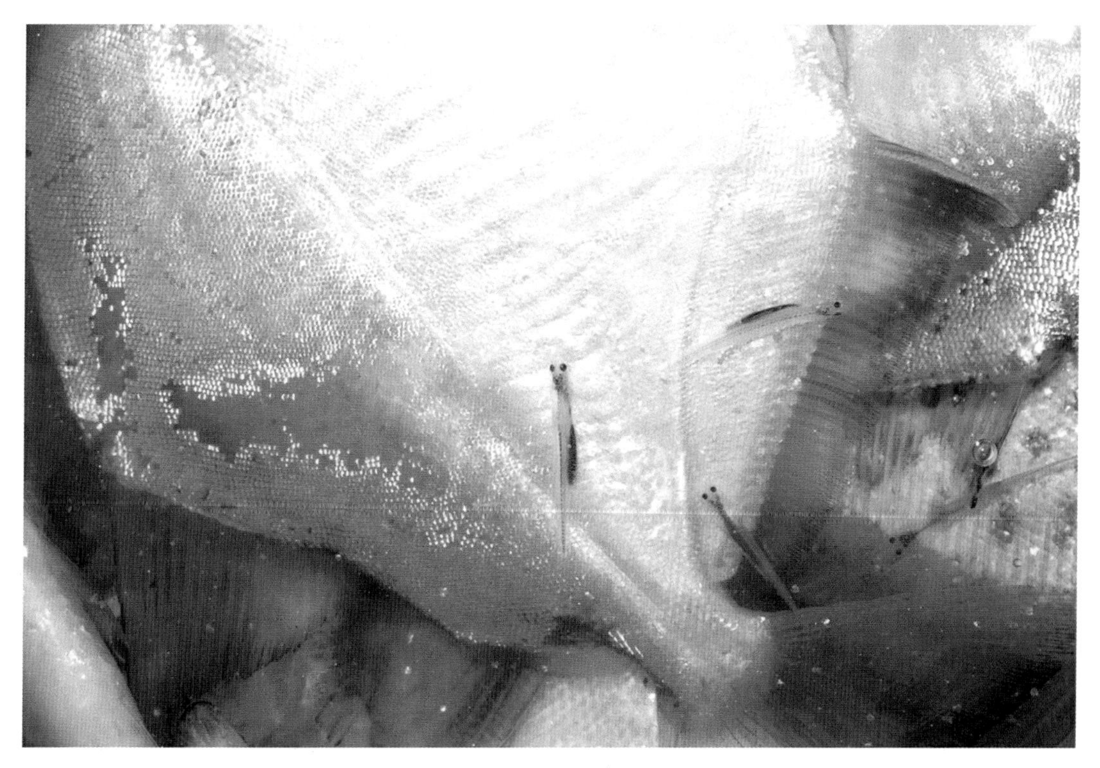

Tiny candirú on a fish captured in a gillnet (bottom photo). These parasitic fish enter the gill chambers of peacock bass and other fish where they feed on blood. *Photo: C. Montaña.*

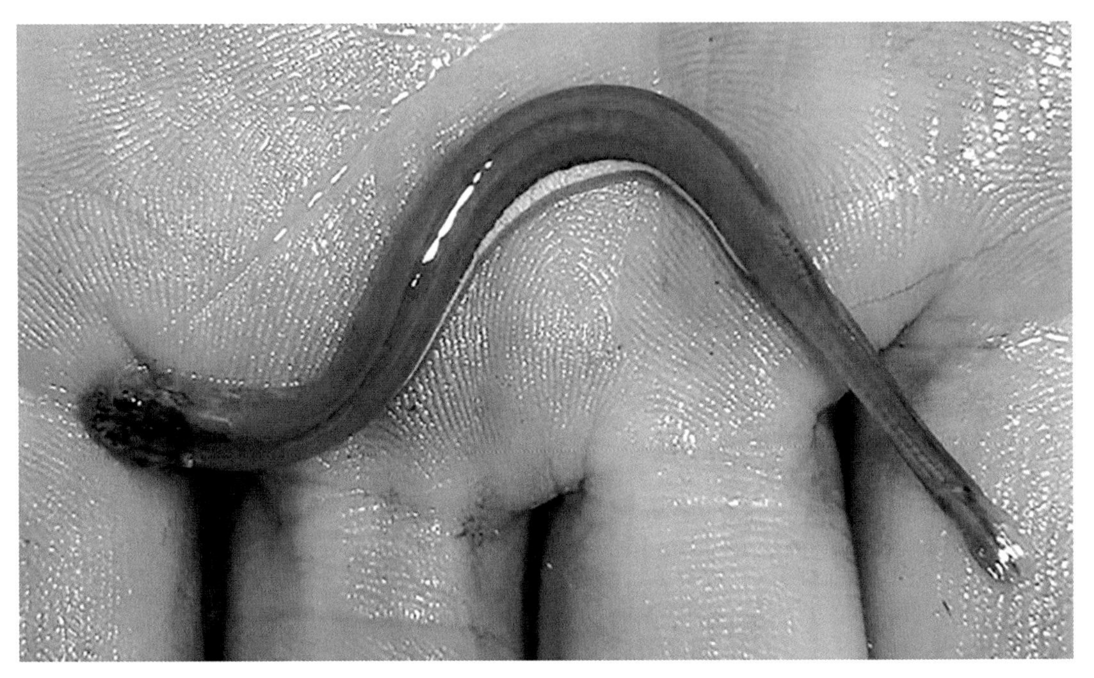

Candirú (*Vandellia* sp.) from the Rupununi River, Guyana. *Photo: K. Winemiller.*

A larger species of candirú, *Vandellia cirrhosa*, gorged with blood. *Photo: C. Montaña.*

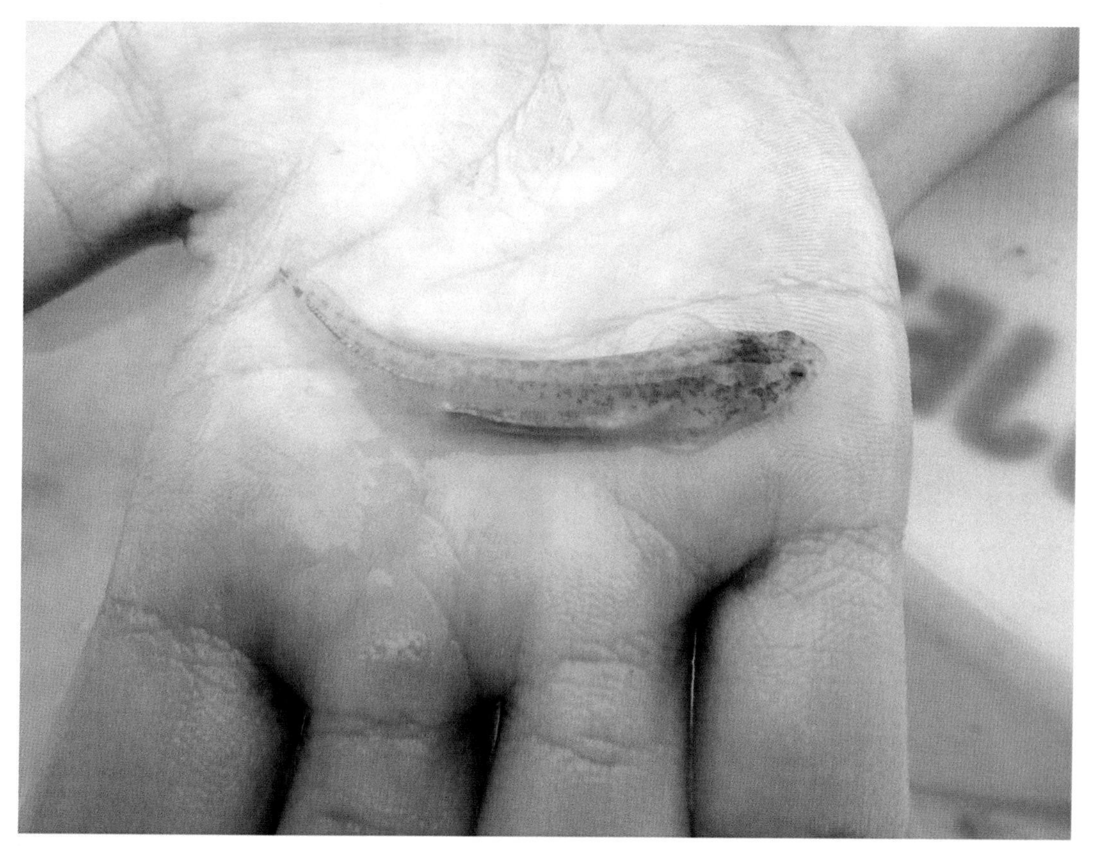

Ochmacanthus alternus, a trichomycterid catfish that feeds only on the mucus covering the bodies of fish. *Photo: K. Winemiller.*

Traits useful for identification of peacock bass species

Identification of peacock bass species is a long-standing problem that is a consequence of continuing discovery of new species and disagreement among ichthyologists. The species accounts that appear in the following chapters provide descriptions of coloration patterns that can be used to identify the nine named species that we consider valid, including one species recently described from the Essequibo Basin in Guyana. We do not attempt to develop a taxonomic key based on morphological traits, such as number of scales in the lateral line and fin ray counts, and that is because some of the species cannot be distinguished based on these features, and genetics data and coloration patterns seem to be more reliable means of identifying species. Obviously, most people will not have access to genetics analyses when catching peacock bass or purchasing fish from ornamental fish suppliers. Most people are not prepared, especially in the field, to examine fin rays and count tiny scales on various parts of the fish's anatomy, and as previously mentioned, those traits are not reliably diagnostic for species identification anyways. Therefore we stress coloration patterns that should allow anyone to identify the nine recognized species fairly quickly. But species identification based on these coloration patterns can be tricky, and that's because of the large variation exhibited by fish of different ages, reproductive states, health status, and even moods (behavioral states) as well as conditions of their habitat.

Basic morphological features involved in descriptions for identification of peacock bass species. *Photo: M. Sabaj.*

Background coloration

Background coloration of the head and body is highly variable both among and within *Cichla* species. Juveniles usually are light gray but sometimes have a yellow tint. Adults of most species can vary from gray to tan to yellow to bronze to green, and a couple of species (*C. piquiti*, *C. intermedia*) sometimes even take on a blue tint. Given the large variation observed within species, background coloration is not a very reliable trait

Variation in the background coloration and markings of the speckled peacock bass (*C. temensis*). *Photos: K. Winemiller.*

for identifying species, although it can be useful when used in combination with other traits. For example, both *C. pinima* and *C. ocellaris* var. *kelberi* are sometimes called yellow peacock bass, and indeed they often have yellow background coloration. However, this is not always the case for either of these species, and large individuals may be gray, bronze, or green.

In all species, the background coloration of the head and body is sometimes covered with a dense pattern of white, beige, or yellow spots or vermiculation (wormy pattern). This seems to be observed more frequently among fishes taken from extremely clear water. Some species, including *C. ocellaris*, *C. orinocensis* and *C. intermedia*, sometimes have tiny black spots all over the body, including dorsal and posterior portions of the head and the caudal peduncle. Tiny black dots are arranged in dense rows that contrast with bright yellow and orange scales to create a strikingly beautiful appearance in the fire peacock bass, *C. mirianae*, from the Rio São Benedito (Upper Tapajos system in Brazil). In the Xingu peacock, *C. melaniae*, these spots tend to be irregular in their shape (blotches), size, and distribution on the body. In rare instances, the skin on the fish's head, body, and/or fins may have large black patches, sometimes covering almost the entire body. This hyperpigmentation is called melanosis, and its causes have not been determined conclusively but could be affected by a viral infection, stress, poor water quality, aging, and/or a genetic condition.

Fire peacock bass (*C. mirianae*) specimen showing rows of tiny black spots against a bright yellow background. This fish has two caudal ocelli, a condition not unusual in other peacock bass species. *Photo: K. Winemiller.*

This peacock bass exhibits extensive melanosis, black pigmentation in the skin. Causes for this uncommon condition are not well understood.

Vertical dark bars

Vertical bars ranging from a barely perceptible shading of the background body color to gray or black are observed in most peacock bass species, especially *C. ocellaris*, *C. temensis*, *C. pinima*, *C. piquiti*, *C. melaniae*, *C. intermedia*, but these bars vary in number and intensity. Bars can fade out and reappear under appropriate conditions (see description of *C. temensis* in Chapter 5). These vertical bars normally are absent or are very faint in *C. mirianae* and *C. orinocensis* and tend to be most apparent in relatively small individuals.

Ichthyologists employ a scheme for describing locations of vertical bars in *Cichla*. One, a few, or all of these elements may be expressed, and some species may show the entire range of bar expression depending on the fish's age, size, physiological state, and environmental conditions. The most commonly expressed bars are 1, 2, and 3, but sometimes there is virtually no expression of vertical bars. *Cichla piquiti* always shows all five vertical bars that extend from the dorsal midline where they are broad toward the belly region where they narrow. *Cichla melaniae* and *C. intermedia* commonly show from 8 to 10 bars of varying width and intensity. The vertical bars sometimes are very dark in fishes that are emaciated or diseased as well as many peacock bass housed in aquaria or inhabiting canals or artificial lakes with extremely clear water.

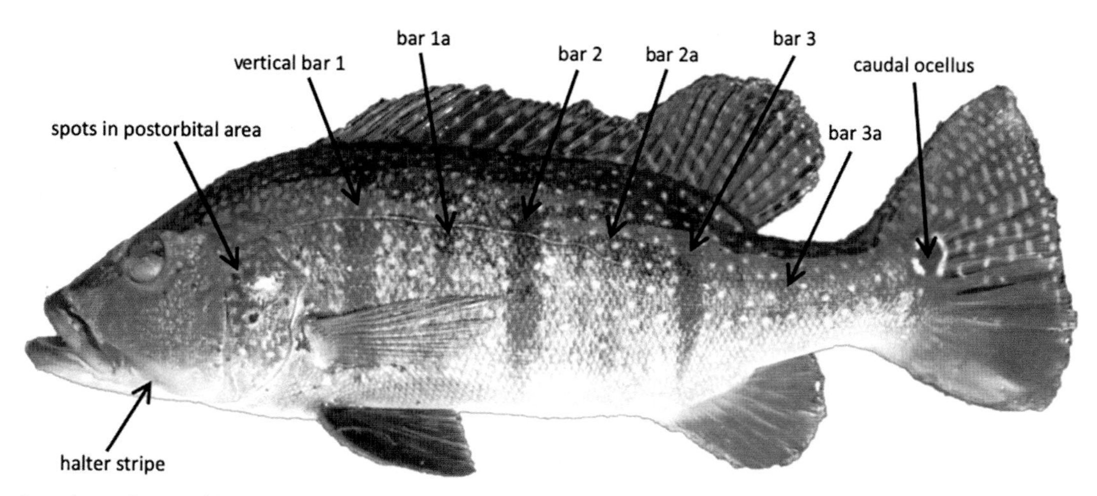

Locations of vertical bars labeled to indicate major bars (1,2,3) and minor bars (1a,2a,3a). Six additional minor bars (faint), black spots in the postorbital area, and profusion of small white spots on the body and fins also can be seen in this specimen of the Xingu peacock bass (*C. melaniae*). *Photo: K. Winemiller.*

In some species, especially *C. ocellaris*, there may be the expression of an occipital bar, a dark diagonal band, on the forehead. The occipital bar can fade in and out, and it seems to be darkest in pre- and postspawning adults and fish from clear water. Some peacock bass may show a so-called halter stripe, dark pigmentation running from the corner of the jaw toward the lower margin of the preoperculum (bone covering the cheek area). A small halter stripe is sometimes present as a darkly shaded area near the corner of the mouth, or this area may have a bright sheen of yellow, green, or blue depending on the species.

Lateral band (stripe)

Small juveniles of all peacock bass have a dark lateral stripe, some complete (e.g., *C. temensis*, *C. pinima*) and some broken and mostly developed in the posterior part of the flank (side of the body) (e.g., *C. ocellaris*, *C. orinocensis*, *C. cataractae* sp.). In most peacock bass species, this lateral band disappears as the fish grows into the adult size class, but in a few there often remains a dark lateral band that often is broken into a series of blotches (*C. intermedia*, *C. mirianae*). In many species, sometimes a series of smaller black blotches form an interrupted lateral band in the caudal peduncle region.

Juvenile *C. orinocensis* (top) and *C. temensis* (bottom) from the lower Branco River. Note the difference in the development of the lateral (horizontal) stripe in these two species. *Photo: K. Winemiller.*

Lateral blotches and ocellated markings

All peacock bass, except for very small juveniles (fish <10 cm length), have a distinct ocellated spot (black spot or blotch surrounded by a bright white or yellow ring) where the caudal peduncle gives rise to the caudal fin. The ocellus is located slightly dorsal to the body midline, and some individuals may have two or three ocelli, a condition sometimes observed in oscars (*Astronotus ocellatus*, Cichlidae), red drum (*Sciaenops ocellatus*, Sciaenidae), and other fishes that have a caudal ocellus. The caudal ocellus is of a size, shape, and often coloration that resembles the fish's eye, and has been proposed to function as an "eye mimic," possibly to deter attacks by small fin-nipping piranhas (*Serrasalmus* spp.) that might confuse the predator's tail end with its head [25, 26]. Another idea is that the caudal ocellus is a signal for species recognition that allows parents to avoid cannibalizing their own young [27]. The problem with that theory is that the caudal ocellus is not expressed until the fry grow to a certain size, and studies of gut contents have revealed cannibalism to be common in most peacock bass populations.

Species that often show ocellated blotches of varying sizes along the body flanks are *C. ocellaris*, *C. orinocensis*, *C. pinima*, *C. mirianae*, and *C. cataractae*. These blotches may be large and completely ocellated with a consistent number—usually three that are located in the areas normally occupied by vertical bands 1, 2, and 3 (e.g., *C. orinocensis*, *C. mirianae*). In other species (e.g., *C. pinima*, *C. cataractae*), these blotches may vary considerably in number, size, degree of ocellation, and location on the body. In some cases, there are two or three ocellated blotches aligned vertically in the positions of vertical bars 1, 2, and 3. In the case of *C. ocellaris* and *C. orinocensis*, there frequently are three lateral blotches, usually ocellated, with rather diffuse vertical bands merging into them from above and below.

Cichla ocellaris, including all of the regional variants throughout the species' extensive geographic distribution, almost always show black abdominal blotches. These blotches have an irregular shape and often run together to form a horizontal band. This pattern is only observed in *C. ocellaris*. *Cichla pinima* sometimes may have dark pigmentation in the abdominal region, but this pattern is usually associated with a lower extension of vertical bar 1 and sometimes bars 2 and 3, often with an irregular shape and a degree of ocellation.

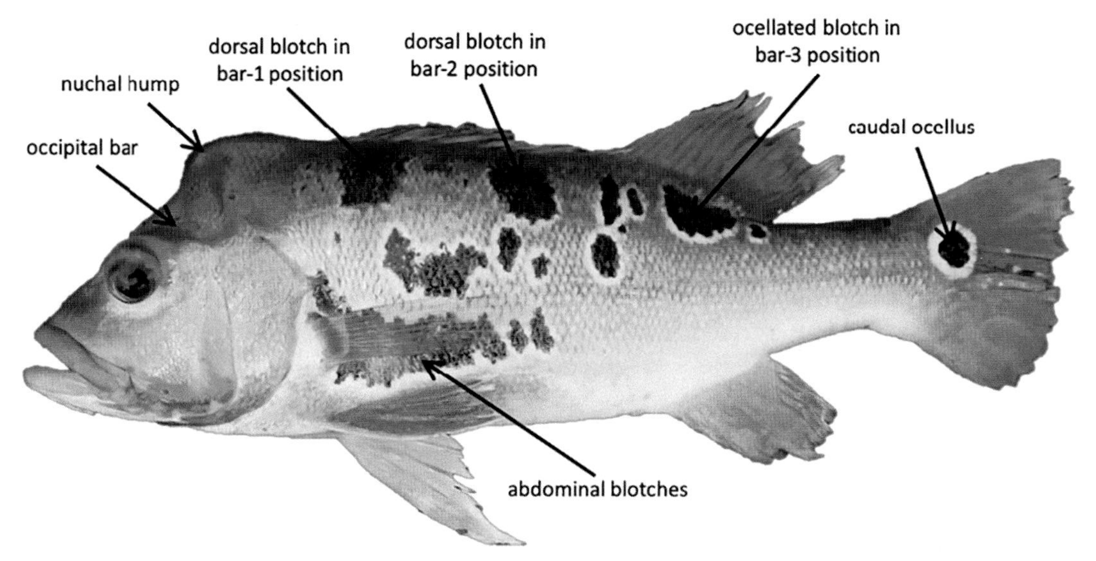

Small male specimen of the butterfly peacock bass (*C. ocellaris*) showing nuchal hump and locations of black blotches characteristic of this species. *Photo: K. Winemiller.*

Postorbital stripe and blotches

Some peacock bass species have black markings on their head that often are surrounded by contrasting white margins in a manner similar to an ocellus. In species, such as *C. temensis* and *C. pinima*, these blotches often merge into a postorbital stripe, with irregular blotches scattered above and below. We have noticed that female *C. temensis* often have more extensive postorbital stripes and blotches than males, but this difference requires further study for confirmation. Other peacock bass species may have only few, relatively small ocellated blotches in the postorbital area of the head (*C. piquiti*), or may have small black spots in this area, either scattered or loosely aligned in the position of a postorbital stripe (*C. melaniae*, *C. mirianae*, *C. intermedia*, *C. cataractae*). Still other species (*C. ocellaris*, *C. orinocensis*) rarely show any black spots on the head.

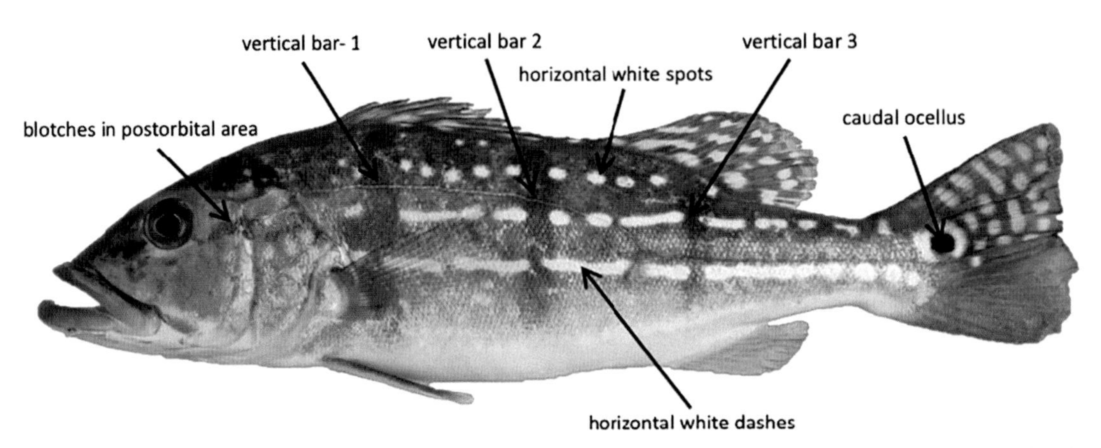

Immature specimen of the speckled peacock bass (*C. temensis*) showing blotches in the postorbital area of the head, three vertical bars on the body, and horizontal rows of white spots and dashes. *Photo: H. López-Fernández.*

Fin coloration

In adult fish, the caudal fin is usually dark on the top half and yellow, orange, or red in the bottom half. In *C. piquiti*, some large breeding individuals are light blue in the lower half of the caudal fin. The top half often has small translucent white or blue spots arranged in irregular rows emanating from the fin base to its posterior margin. The intensity of the color in lower half of the caudal fin varies in relation to water conditions and reproductive state, with most vivid colors observed in prespawning, nesting, and brood-guarding individuals.

The pelvic and anal fins generally have the same color as the lower half of the caudal fin, with intensity varying in the same manner. The pectoral fins essentially are transparent. The dorsal fin is usually gray, often with rows of translucent white or light blue spots, especially on the rayed dorsal fin (posterior portion that lacks spines). In some species, the spinous dorsal fin and rayed dorsal fin can be blue (e.g., *C. piquiti*, *C. intermedia*), and most species may show a bluish or green sheen under certain conditions. In some populations of *C. ocellaris*, there are bright yellow spots on the rayed portion of the dorsal, caudal, anal, and pelvic fins. This feature is most frequently observed among fish in the Tocantins Basin and was the principal diagnostic trait offered by K&F in support of a proposed species status (*C. kelberi*) for those stocks. However, many individuals in that basin lack this trait, and some *C. ocellaris* from other regions display bright yellow spots on these fins.

Despite the fact that peacock bass coloration varies greatly within species, careful examination of a handful of coloration features should allow any person to identify the nine peacock bass species recognized in this book. In the following chapters, we summarize the features unique to each species.

Butterfly peacock bass (*C. ocellaris* var. *kelberi*) caught from a backwater in the Paraná River where it is an introduced species. Note the yellow spots on the pelvic, anal, and caudal fins. Fish from the same population are found with and without these spots. *Photo: K. Winemiller.*

References

[1] S.O. Kullander, E.J.G. Ferreira, A review of the South American cichlid genus *Cichla*, with descriptions of nine new species (Teleostei: Cichlidae), Ichthyol. Explor. Freshw. 17 (2006) 289–398.

[2] S.C. Willis, M.S. Nunes, C.G. Montana, I.P. Farias, N.R. Lovejoy, Systematics, biogeography, and evolution of the neotropical peacock basses *Cichla* (Perciformes: Cichlidae), Mol. Phylogenet. Evol. 44 (1) (2007) 291–307.

[3] S.C. Willis, J. Macrander, I.P. Farias, G. Orti, Simultaneous delimitation of species and quantification of interspecific hybridization in Amazonian peacock cichlids (genus *Cichla*) using multi-locus data, BMC Evol. Biol. 12 (2012) 96, https://doi.org/10.1186/1471-2148-12-96.

[4] S.C. Willis, I.P. Farias, G. Orti, Multi-locus species tree for the Amazonian peacock basses (Cichlidae: *Cichla*): emergent phylogenetic signal despite limited nuclear variation, Mol. Phylogenet. Evol. 69 (3) (2013) 479–490.

[5] D.B. Jepsen, K.O. Winemiller, D.C. Taphorn, Temporal patterns of resource partitioning among *Cichla* species in a Venezuelan blackwater river, J. Fish Biol. 51 (1997) 1085–1108.

[6] K.O. Winemiller, Ecology of peacock cichlids (*Cichla* spp.) in Venezuela, J. Aquaric. Aquat. Sci. 9 (2001) 93–112.

[7] D. Rodriguez-Olarte, D.C. Taphorn, Aspectos de la ecología reproductive del pavón Estrella *Cichla orinocensis* Humboldt 1833 (Pisces:Perciformes:Cichlidae) en el Parque National Aguaro-Guariquito, Venezuela, Mem. Fund. La Salle Cienc. Nat. 161–161 (2005) 5–17.

[8] C.G. Montaña, D.C. Taphorn, C.A. Layman, C. Lasso, Distribución, alimentación and reproducción de tres species de pavones Cichla spp. (Perciformes, Cichlidae) en la cuenca del Rio Ventuari, Venezuela, Mem. Fund. La Salle Cienc. Nat. vol. 165 (2007) 83–102.

[9] H. Muñoz, P.A. VanDamme, F. Duponchelle, Breeding behavior and distribution of the tucunaré *Cichla* aff. *monoculus* in a clear water river of the Bolivian Amazon, J. Fish Biol. 69 (2006) 1018–1030.
[10] S.C. Willis, One species or four? Yes! Or, arbitrary assignment of lineages to species obscures the diversification processes of Neotropical fishes, PLoS One 12 (2) (2017), e0172349.
[11] O. Fontenele, Contribução para o conhecimiento de biologia dos tucanarés, Actinopterygii Cichlidae, em cativiero. Aprelhlo de reproducao. Habitos de desova incubaçã, Rev. Bras. Biol. 10 (1950) 503–519.
[12] J.E. de Souza, E.N. Fragoso-Moura, N. Fenerish-Verani, O. Rocha, J.R. Verani, Population structure and reproductive biology of *Cichla kelberi* (Perciformes, Cichlidae) in Lobo Reservoir, Brazil, Neotrop. Ichthyol. 6 (2) (2008) 201–210.
[13] F.M. Pelicice, A.A. Agostinho, Fish fauna destruction after the introduction of a non-native predator (*Cichla kelberi*) in a neotropical reservoir, Biol. Invasions 11 (2009) 1789–1801.
[14] L.M. Gomeiro, A.L. Carmassi, G.R. Rondineli, G.A. Villares Junior, F.M.S. Braga, Growth and mortality of Cichla spp. (Perciformes, Cichlidae) introduced in Volta Grande Reservoir (Grande River) and in a small artificial lake in Southeastern Brazil, Braz. J. Biol. 70 (4) (2010) 1093–1101.
[15] H. Silva Mendonça, A.C. Alcânara Santos, M. Marques Martins, F. Gerson Araújo, Size-related and seasonal changes in the diet of the non-native *Cichla kelberi* Kullander & Ferreira, 2006 in a lowland reservoir in the southeastern Brazil, Biota Neotrop. 18 (3) (2018), e20170493.
[16] T.M. Zaret, Life-history and growth relationships of *Cichla ocellaris*, a predatory South American cichlid, Biotropica 12 (1980) 144–157.
[17] D.M.T. Sharpe, L.F. De Leon, R. González, M.E. Torchin, Tropical fish community does not recover 45 years after predator introduction, Ecology 98 (2) (2017) 412–424.
[18] J.W. Neal, J.M. Bies, C.N. Fox, C.G. Lilyestrom, Evaluation of proposed speckled peacock bass *Cichla temensis*i to Puerto Rico, N. Am. J. Fish Manag. 37 (2017) 1093–1106.
[19] W.S. Devick, Life history study of the tucunaré *Cichla ocellaris*, in: Federal Aid in Sportfish Restoration Project F-9-1, Job Completion Report, Hawaii Department of Land and Natural Resources, Honolulu, 1972.
[20] P.L. Shafland, An overview of Florida's introduced butterfly peacock bass (*Cichla ocellaris*) sportfishery, Natura 96 (1993) 26–29.
[21] H. Sioli, The Amazon and its main affluents: hydrography, morphology of the river courses, and river types, in: H. Sioli (Ed.), The Amazon, Monographie Biologicae, vol. 56, Dr. W. Junk Publishers, The Hague, Netherlands, 1984, pp. 127–165.
[22] K. Furch, W.J. Junk, The chemical composition, food value, and decomposition of herbaceous plants, leaves, and leaf litter of floodplain forests, in: The Central Amazon Floodplain, Ecological Studies, vol. 126, Springer, 1997, pp. 187–205.
[23] S. Spotte, Candiru: Life and Legend of the Bloodsucking Catfishes, Creative Arts Book Company, Berkeley, CA, 2002. 322 p.
[24] K.O. Winemiller, H.Y. Yan, Obligate mucus feeding in a South American trichomycterid catfish (Pisces: Ostariophysi), Copeia 1989 (1989) 511–514.
[25] K.O. Winemiller, Spatial temporal variation in tropical fish trophic networks, Ecol. Monogr. 60 (1990) 331–367.
[26] K.O. Winemiller, L.C. Kelso-Winemiller, Fin-nipping piranhas, Natl. Geogr. Res. 9 (1993) 344–357.
[27] T.M. Zaret, R.T. Paine, Species introduction in a tropical lake, Science 182 (1973) 449–455.

CHAPTER

2

Butterfly peacock bass, *Cichla ocellaris* (Bloch & Schneider 1801)

Cichla ocellaris

Origin of name

Latin "*ocellaris*" meaning "a little eye," a reference to the prominent caudal ocellus

Synonyms

Cychla monoculus Agassiz, 1831, *Cychla nigro-maculata* Jardine, 1843; *Acharnes speciosus* Müller & Troschel, 1849; *Cycla toucounarai* Castelnau, 1855; *Cichla bilineatus* Nakashima, 1941; *Cichla kelberi* Kullander & Ferreira 2006; *Cichla pleiozona* Kullander & Ferreira, 2006

Common names

C. ocellaris var. *ocellaris* (Guyana)—Lukanani, Pond lukanani, Tucunaré-assu

C. ocellaris var. *ocellaris* (Florida and other English-speaking regions)—Peacock bass, Butterfly peacock bass, Yellow peacock bass

https://doi.org/10.1016/B978-0-323-85157-2.00002-0

C. ocellaris var. *monoculus*, *C. ocellaris* var. *kelberi*, *C. ocellaris* var. *pleiozona* (Brazil)—Tucunaré, Tucunaré mariposa, Tucunaré amarillo

C. ocellaris var. *pleiozona* (Bolivia)—Tucunaré

C. ocellaris var. *monoculus* (Colombia, Ecuador, Peru, Venezuela)—Pavón, Pavón mariposa, Marichapa

Geographic distribution

Cichla ocellaris has the largest geographic distribution among all peacock bass species, occurring throughout nonmountainous regions of the western Amazon, including the Madeira Basin in Bolivia, Brazil, and Peru. The species also is common along the length of the Amazon to its mouth as well as throughout the Negro Basin, Upper Orinoco, Tocantins, and other lower Amazon tributaries. In the Xingu River, *C. ocellaris* var. *monoculus* is only found in the lowest reach below the cataracts in the Volta Grande to the river's confluence with the Amazon. The species also is distributed in rivers flowing to the Atlantic Ocean and Caribbean Sea in northeast Brazil, French Guiana, Suriname, and Guyana.

Natural history

Tolerates a greater range of water quality than other peacock bass, which may account for its large geographic distribution compared to other species. Widely distributed in side channels and lakes within the vast floodplains of the Amazon River. Occupies shallow water close to shore with complex submerged structure. Primarily piscivorous.

A few years ago, two of us (KOW, CGM) had the opportunity to join a scientific expedition to the Rupununi region of Guyana to survey fishes. We were eager for the field trip, because this was a region of South America that neither of us had yet explored, and we would have an opportunity to catch *Cichla ocellaris*, the type species for the genus (the first valid species name assigned to the genus *Cichla*). *Cichla ocellaris* has special significance not only for ichthyologists, but also for fisheries biologists and anglers, because it has been introduced and thrives in various locations scattered across the globe. By the time of the expedition to Guyana, fortuitously the first of several for us (see Chapter 10), we had thoroughly read and absorbed K&F's 2006 monograph containing a complete revision of *Cichla* taxonomy. Sven Kullander is a Swedish ichthyologist and leading expert on Neotropical cichlid taxonomy, and Efrem Ferreira, one the most important ecologists working on Amazon fishes, is from Brazil's Instituto Nacional de Pesquisas do Amazonia (National Institute for Amazonian Research). In this landmark study, the authors attempted to sort out the complicated state of *Cichla* taxonomy, and in the process, they described nine new species. As we explain in this book, subsequent genetic analyses [1–3] determined that four of those species appear to be perfectly valid, but five of them are more accurately characterized as local varieties that lack unique differences in morphology, coloration pattern or, most importantly, genetic composition. Two of K&F's new species (*C. kelberi*, *C. pleiozona*) as well as two older names (*C. monoculus*, *C. nigromaculata*) are not genetically distinct from *C. ocellaris*, the oldest name that is valid for this species.

We had previously collected all *C. ocellaris* varieties except for the originally described form that is found in Guyana, Suriname, and French Guiana, and at last we had a chance to observe fish from that region. Many years ago, KW visited some French colleagues working in Bolivia's Mamoré River (Upper Madeira Basin) and observed a peacock bass they had caught in an experimental gillnet. That fish looked identical to a peacock bass we had captured in the Casiquiare River in southern Venezuela. Several years later, KW was with Brazilian colleagues conducting research on the Upper Paraná River and had the opportunity to fish for *C. kelberi* that had been translocated from the Tocantins Basin and stocked into Upper Paraná reservoirs. Those fish looked similar to the peacock bass observed in Bolivia and southern Venezuela. All of these peacock bass resembled the *C. monoculus* that KW observed during multiple visits to the Amazon and Rio Negro. However, the year was 2009, and the genetics studies had not yet been published, therefore the best information about *Cichla* taxonomy was contained in K&F's impressive monograph.

There are two ways to travel from Georgetown, the Guyanese capitol city located on the coast, to the Rupununi Savanna District in the country's interior—the easy route by air, and the difficult route by road. Naturally, we took the difficult route. We loaded two 4-wheel-drive vehicles with camping gear, instruments for measuring environmental parameters, fish collecting equipment, preservation chemicals and containers, and photographic supplies. The road from Georgetown to the city of Linden is paved, but later it turns to dirt riddled with muddy potholes and ruts, creating a tortuous path that meanders over hills and valleys to the town of Lethem in the Rupununi Savanna District near the border with Brazil. To say the least, the journey was uncomfortable, but after taking the ferry across the Essequibo River, we made our way to the Pirara River, a small tributary of the Takutu River, which in turn is a tributary of the Branco, a great clearwater river that flows to the Rio Negro in the Brazilian Amazon.

Male *C. ocellaris* showing nuchal hump and bright red iris characteristic of nesting and brood-guarding fish. This fish was caught in the Rupununi River by local naturalist and guide Ashley Holland. *Photo: A. Holland.*

We hung our hammocks under the roof overhang of a dilapidated building that had seen better days when the property was a working ranch, now long since abandoned. It was safer to work, eat, and sleep outside the building, because inside, the walls were plastered with bat guano and adorned with a massive wasp nest. The wasps were swarming and already one of our team had been stung. We began setting gillnets and pulling a large seine net, and soon discovered that the Pirara contained impressive fish abundance and diversity. A headwater of the Branco River, the Pirara lies within the Amazon Basin and not the Essequibo Basin. Yet all available reports claim *C. ocellaris* inhabits rivers throughout the Rupununi Savanna, and since this area floods extensively during the wet season, most kinds of fish are able to disperse between the Rupununi River (Essequibo) and Upper Branco tributaries such as the Pirara. Using rod and reel, I caught two small *C. temensis*, but not a single *C. ocellaris*. Fortunately, my colleague Ryan King, aquatic ecologist from Baylor University, brought fly-fishing gear, and the next morning he managed to land five adult *C. ocellaris* from a single pool.

Floodwaters on the Rupununi Savanna of southern Guyana. During the rainy season, this region is a potential corridor for fishes to disperse between the headwaters of the Branco and Essequibo rivers. *Photo: C. Montaña.*

Immature *C. ocellaris* from the Pirara River, Guyana. *Photo: H. López-Fernández.*

These fish closely resembled most of the fish identified by K&F as *C. monoculus, C. nigromaculata, C. kelberi,* and *C. pleiozona*. Like most *Cichla* species, all of the fish represented by these species' names show a broad range of color patterns, yet always with some consistent features. No morphological or coloration characteristics have yet been discovered that can reliably distinguish fish from these five regional stocks from one another. For example, the only unique attribute proposed to distinguish *C. kelberi* from other *Cichla* species is the presence of yellow spots on the anal and pelvic fins. However, not all *C. kelberi* from the Tocantins River and locations in the Upper Paraná where this stock has been introduced have this characteristic, and sometimes fish from Bolivia or the Amazon have some yellow spots on the anal and pelvic fins. Similarly, many people classify any *C. ocellaris* showing an occipital bar (diagonal dark stripe on the forehead) as *C. pleiozona*, the stock native to the Upper Madeira in Bolivia, southern Brazil, and southern Peru, but fish from other regions also express this feature under certain conditions. During a recent visit to the ichthyology collection of the Museu Emilio Goeldi in Belem, Brazil, KW found specimens collected from the region of the Amazon estuary that were classified as *C. ocellaris, C. kelberi, C. monoculus, C. nigromaculata,* and *C. pleiozona*. Sometimes two or more of these were collected from the same location. The most likely explanation is that the identifiers had relied on coloration patterns emphasized in K&F's monograph. However, those color patterns turn out to be highly variable in fish from all five regions. It is highly unlikely that someone transported fish from the Upper Madeira, Guyana, and Casiquiare regions to the Amazon estuary, and that those populations subsequently thrived there without interbreeding. Given the new genetic findings, it seems prudent to refer to these stocks as regional varieties of *C. ocellaris*. These varieties possess certain differences in their genetic composition that belie various degrees of evolutionary divergence over time and space (see Chapter 11), but the genetic data also reveal extensive mixing of gene pools. Essentially, the choice to designate these regional stocks as separate "species" versus "varieties or subspecies" of *C. ocellaris* depends on which of several species concepts one chooses to adopt, plus how one decides to weigh evidence from analyses of morphology, coloration, and genetics. Since this book is primarily aimed at readers who are not professional scientists, we rely heavily on coloration patterns for species identification. The species we recognize based on coloration patterns agree well with species designations based on recent genetics analyses [1–3].

Rewa River (Essequibo)

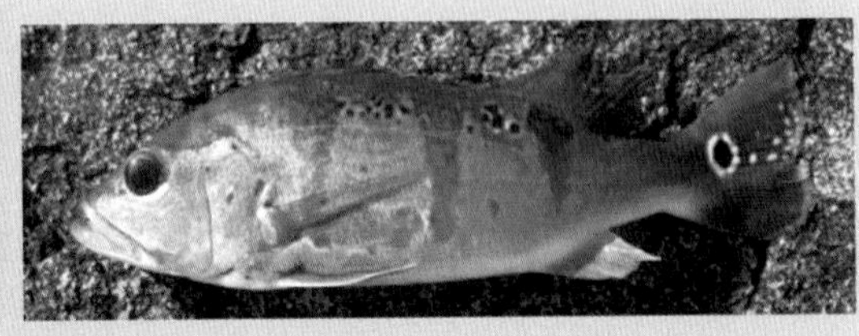
Pamoni River (Casiquiare, Amazon-Orinoco)

Rio Negro (Amazon)

Lower Branco River (Amazon)

Jari River (Amazon)

Jacundá River (Amazon)

Lower Xingu River (Amazon)

Juruá River (Amazon)

Lajeado Reservoir (Tocantins)

Lajeado Reservoir (Tocantins)

Upper Paraná River (introduced from Tocantins)

Identification

Cichla ocellaris was first described by Bloch & Schneider (1801) based on a color drawing of a specimen of uncertain origin (recorded only as "India Orientali"). The specific origin of the designated type specimen also is unknown [4]. For centuries, all peacock bass were identified only as *C. ocellaris*, but today we know there are at least nine distinct species, one of which was described while we were writing this book (see Chapter 10). Depending on different interpretations of morphological and genetic variation and species concepts, there could be 17 or more *Cichla* species. Various lines of evidence and opinions about *Cichla* taxonomy will be reviewed in Chapter 11, and here we consider *C. kelberi*, *C. monoculus*, *C. nigromaculata*, and *C. pleiozona* to be regional stocks that lack reliable morphological and coloration characteristics that can uniquely separate one from another. At the same time, these regional stocks have been shown to have certain genetic differences that suggest varying degrees of geographic isolation, as well as other genetic evidence indicating recent natural interbreeding among regional stocks.

German naturalist Marcus Elieser Bloch's illustration of *Cichla ocellaris* from the 1801 species description.

The following features pertain to all varieties of *Cichla ocellaris*. The background coloration of the head and body of adult fish varies from grayish-green to olive-green to yellow, with the dorsal region darker than the flanks, and the ventral region (belly) usually white. The background coloration and intensity of markings of an individual fish can change in a matter of minutes after capture, for example green may fade to yellow. In large individuals (> 20 cm), three vertical bars extend from the dorsal region onto the flanks where they gradually narrow at their tips. The third bar typically has an ocellated blotch located above the body midline. In larger adults (> 30 cm), the dorsal portion of bars 1 and 2 also has dark blotches that are sometimes ocellated but most often are not. Sometimes only the dorsal region of these vertical bars is expressed.

Subadult and adult *C. ocellaris* lack a postorbital stripe and blotches in the cheek area. With rare exceptions, even small black spots are completely absent from the head. Adults and many subadults have a black, abdominal blotch, a trait that distinguishes *C. ocellaris* from all other peacock bass species. Expression of the abdominal blotch is highly variable, ranging from a few relatively small blotches, to a large band of blotches that extend horizontally with smaller blotches scattered across the lower flanks from near the base of the pectoral fin to the base of the anal fin. In subadults and even some adults, the abdominal blotch may be poorly developed, and yet some degree of dark pigmentation can be seen on the abdomen in the area below the pectoral fin. Some populations of *C. ocellaris* var. *monoculus*, such as those in the Rio Negro Basin, show extensive blotches scattered across the entire body, with some of them partially or fully ocellated.

A breeding pair of *C. ocellaris* var. *monoculus* taken from a backwater in the lower Branco River, Brazil. These fish are unusual in having multiple large, ocellated blotches on the body. Note that only the dorsal portion of the three vertical bars is expressed in these nesting fish. *Photo: K. Winemiller.*

Most *C. ocellaris* populations have at least some individuals with small black spots scattered across the sides of the body, especially the dorsal region. These small black spots are supposed to be a distinguishing characteristic for *C. nigromaculata* (which, based on findings from genetic studies, we do not recognize as a separate variety distinct from *C. ocellatus* var.

monoculus); however, they can be found in specimens of all *C. ocellaris* populations. Some individuals may have their head, body, and fins covered with cloudy white or beige spots, and breeding fish in some populations in coastal rivers of Guyana and Suriname sometimes have rows of tiny black spots all over the body. Rarely observed are specimens with yellow wormy pigmentation covering the head and body. These last two patterns are frequently observed among fishes inhabiting clear water in the Araguaia-Tocantins basin (*C. ocellaris* var. *kelberi*), Brokopondo Reservoir in Suriname, Lake Gatun in Panama, and canals of south Florida.

Unusual black spotting pattern of a breeding *C. ocellaris* taken from the New River, Guyana. This fish also reveals large black spots dorsally where the three vertical bars normally appear. *Photo: M. Charles.*

The dorsal fin and upper half of the caudal fin vary from light gray to dark gray, and these fins may have a light blue or blue-green tint, streaking or spotting. The ventral fins, anal fin, and lower half of the caudal fin generally are orange, orange-red, or red, with intensity greatest during the reproductive period. The caudal fin, rayed portion of the dorsal fin, pelvic fins, and anal fin may have variable numbers of small yellow or white spots. Fish from the Araguaia-Tocantins Basin (*C. ocellaris* var. *kelberi*) appear to be more prone to displaying large numbers of yellow spots on these fins, but specimens with few spots or completely lacking these spots can be found in the same population as fish with heavily spotted fins.

C. ocellaris var. *kelberi*, Porto Rico, Upper Paraná River. Note this individual has a distinct occipital bar and wormy yellow spots on pelvic and anal fins. *Photo: K. Winemiller.*

Specimen of *C. ocellaris* var. *kelberi* from Lajeado Reservoir (Tocantins River) lacking expression of vertical bars or spots on the body or development of yellow spotting on pelvic and anal fins. *Photo: K. Winemiller.*

Cichla ocellaris var. *monoculus* from the Roosevelt River (Madeira Basin). Note yellow spotting on pelvic and anal fins, which was proposed as a diagnostic characteristic in the description of *C. kelberi*, a lineage restricted to the Tocantins Basin. This vividly colored fish also shows a very distinct occipital bar. *Photo: S. Willis.*

During breeding season, individuals of both sexes develop bright orange-red coloration in the region of the throat and sometimes extending onto the belly region, pelvic fins, anal fin, and lower half of the caudal fin. As observed in all peacock bass species, male *C. ocellaris* form a nuchal hump just prior to and during the reproductive period, and the iris of the eyes of both sexes turns brilliant red.

Juveniles smaller than 5 cm are silver-gray with a white belly, and they later develop the background coloration described for adults. Distinct vertical bars, blotches, and the caudal ocellus, the latter being characteristic of all peacock bass, are absent in juveniles smaller than 5 cm. Juveniles have three black spots on the flanks in the positions where vertical bars 1, 2, and 3 develop later. Sometimes there is narrow vertical shading occurring above and below these spots. In juveniles, a thin, horizontal black stripe runs from the third spot to the tip of the caudal peduncle, a pattern that normally disappears as the fish grows. Individuals larger than 16 cm often have an ocellated blotch dorsally in the position of vertical bar 3. The fins of small juveniles are transparent, and as the fish grow, their dorsal, caudal, and anal fins turn to light gray with white spots.

Cichla ocellaris var. *monoculus* from the lower Rio Negro, Brazil. The abdominal blotch is well developed in this specimen. *Photo: K. Winemiller.*

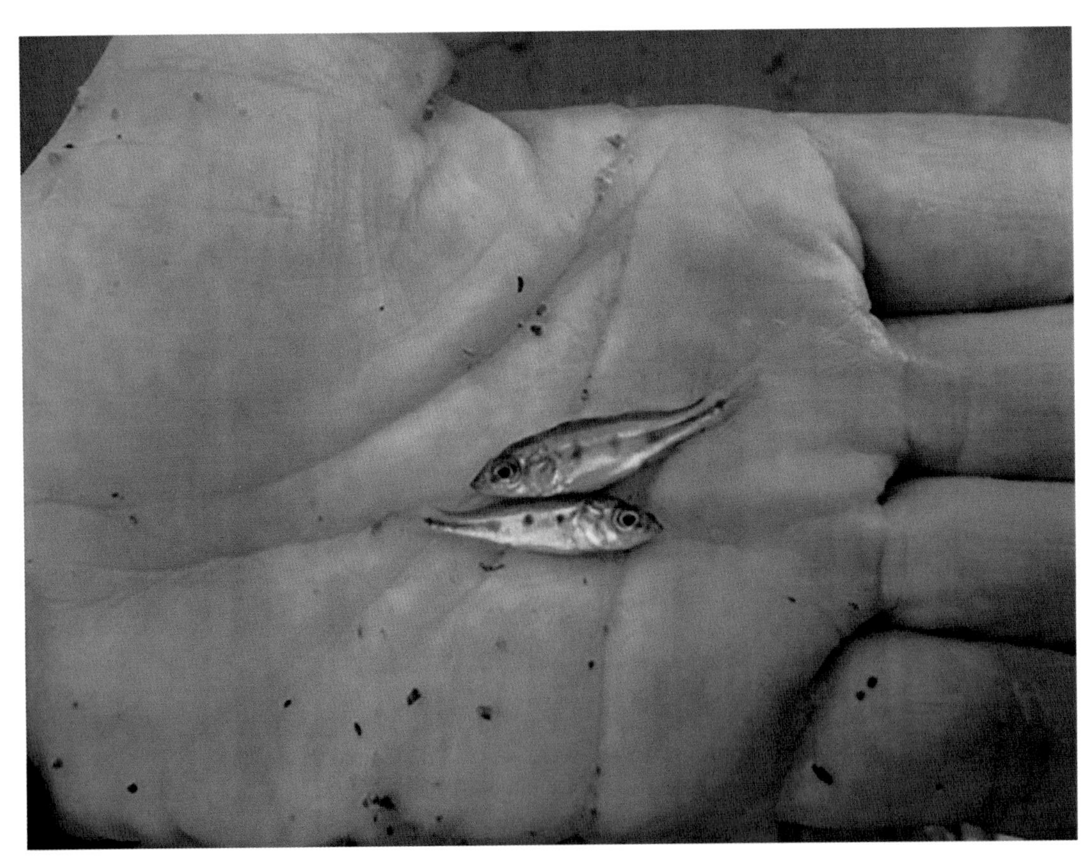

Juvenile *C. ocellaris* from the Rewa River, Guyana. *Photo: K. Winemiller.*

Juvenile *C. ocellaris* var. *monoculus* from Colombia obtained in the United States from a commercial importer. This fish had fed greedily on mosquitofish prior to being photographed. *Photo: K. Winemiller.*

Immature *C. ocellaris* var. *kelberi* from Lajeado Reservoir (Tocantins River). *Photo: K. Winemiller.*

Distribution and habitat

As mentioned previously, *Cichla ocellaris* has at least four geographic varieties (sometimes referred to as "stocks" or "subspecies") that we consider sufficiently distinct genetically to warrant recognition. As discussed before, K&F considered these varieties to be separate species, but recent genetic research has shown high similarity among these groups [1, 2]. One of the previously described species considered as valid by K&F, *C. nigromaculata*, is virtually indistinguishable from *C. ocellaris* var. *monoculus* from the Rio Negro, and there are no geographic barriers to prevent mixing of gene pools. Therefore we consider these to be a single variety, *C. ocellaris* var. *monoculus*. *Cichla nigromaculata* was described based on a handful of specimens from the Casiquiare River and Upper Orinoco in southernmost Venezuela. The Casiquiare originates as a channel from the Upper Orinoco and then flows to the Upper Rio Negro on the border between Colombia and Venezuela just upstream from both of these countries' borders with Brazil. Thus the Casiquiare is a unique river that joins together the great Orinoco and Amazon basins.

Cichla ocellaris var. *monoculus* specimen captured from a lake in the floodplain of the lower Branco River. This fish has a pigmentation pattern consistent with the description of *C. nigromaculata* which was supposed to be restricted to the upper Rio Negro and Casiquiare rivers, and which we consider to be a synonym of *C. ocellaris* var. *monoculus*. *Photo: K. Winemiller.*

If we consider *C. ocellaris* to be a single species, it has the largest geographic distribution among all peacock bass species. *Cichla ocellaris* occurs throughout nonmountainous regions of the western Amazon, including the Madeira Basin in Bolivia, Brazil, and Peru. The species also is common in lowland habitats along the length of the Amazon to its mouth as well as throughout the Negro Basin, Upper Orinoco, Tocantins, and other lower Amazon tributaries in reaches below major cataracts and rapids that apparently function as barriers to upstream dispersal. In the Xingu River, *C. ocellaris* var. *monoculus* is only found in the lowest reach below the cataracts in the Volta Grande (Big Bend) until the river's confluence with the Amazon. A different species, *Cichla melaniae* (Chapter 8) is found in the river's swiftly flowing reaches upstream [5–7]. *Cichla ocellaris* also occurs naturally in rivers flowing to the Atlantic Ocean and Caribbean Sea in northeast Brazil, French Guiana, Suriname, and Guyana.

A sample of *C. ocellaris* var. *monoculus* from the lower Xingu River in the reach below the last rapids and its confluence with the Amazon River. *Photo: S. Willis.*

Cichla ocellaris var. *monoculus* is native to lowland reaches of rivers and lakes throughout the Amazon Basin from Peru, Ecuador, and Colombia along the Amazon mainstem to the Atlantic coast, including coastal drainages in northeast Brazil (e.g., Araguari River) and French Guiana (e.g., Oyapock River). *Cichla ocellaris* var. *monoculus* also is common throughout the Rio Negro Basin and into the Casiquiare and Upper Orinoco.

Upper Orinoco River with the Cerro Duida tepui in the background. This region supports stocks of *C. ocellaris* var. *monoculus* (formerly recognized as a separate species, *C. nigromaculata*, that recently was shown to be genetically indistinct from *C. ocellaris* var. *monoculus*). *Photo: K. Winemiller.*

C. ocellaris var. *monoculus* from the Emoni River, a tributary of the Siapa River, a major clearwater tributary of the Casiquiare in southern Venezuela. *Photo: K. Winemiller.*

Cichla ocellaris var. *kelberi* is native to the Tocantins-Araguaia Basin in Brazil and has been stocked in lakes and reservoirs throughout the Upper Paraná Basin where peacock bass are not native. The Tocantins River empties into a channel that carries water from the Amazon River to the Atlantic coast just south of the huge Marajó Island near the city of Belem. The lower Tocantins has a series of rapids that apparently were a barrier that limited dispersal between populations of *C. ocellaris* upstream (*C. ocellaris* var. *kelberi*) from those downstream and in the lower Amazon (*C. ocellaris* var. *monoculus*). Based on genetic analysis, there is insufficient evidence to consider *C. ocellaris* var. *kelberi* to be a distinct species [1]. Moreover, there are no coloration or morphological characteristics that can reliably separate *C. ocellaris* var. *kelberi* from other varieties of *C. ocellaris*. The principal diagnostic characteristic was given as the presence of gold or yellow spots on the anal and pelvic fins. However, we have seen fish with and without these yellow spots on the fins within the native range of *C. ocellaris* var. *kelberi* as well as in the Upper Paraná where this fish was introduced. Also, some specimens among each of the other *C. ocellaris* varieties have yellow spots on the anal, pelvic, soft dorsal, and anal fins.

Cichla ocellaris var. *pleiozona* occurs in the Upper Madeira Basin in Peru, Bolivia, and Brazil. A series of rapids in the Middle Madeira River near the city of Porto Velho, Brazil, apparently

Geographic distribution of *Cichla ocellaris*. Genetically divergent varieties are shown in different colors: green = var. *ocellaris*, red = var. *monoculus*, blue = var. *pleiozona*, and yellow = var. *kelberi*. Purple shaded area is where *C. ocellaris* var. *kelberi* has been introduced.

limited dispersal between populations of *C. ocellaris* var. *pleiozona* and *C. ocellaris* var. *monoculus* downstream. The coloration and morphological characteristics on which K&F's description of *C. pleiozona* was based cannot be used to separate these fish from other varieties of *C. ocellaris*; moreover, fish from the Upper Madeira lack sufficient genetic divergence to be considered a distinct species. It was proposed that *C. pleiozona* is distinguished, in part, by the presence of an occipital bar and vertical bar 4 on the dorsal portion of the caudal peduncle, yet many specimens lack these pigmentation patterns, and these features can fade in and out in all varieties of *C. ocellaris*.

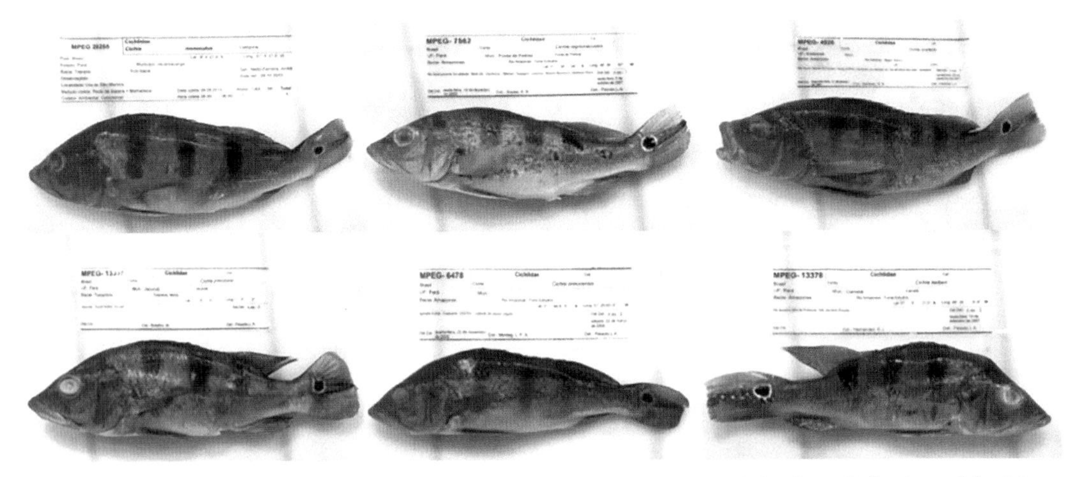

Preserved specimens of immature *C. ocellaris* var. *monoculus* archived in the Ichthyology Collection of the Museu Emilio Goeldi in Belém, Brazil. Only the top-left specimen was correctly identified; the others were erroneously identified as species that would be far outside their presumptive geographic distributions: *C. nigromaculata* from the Amazon estuary (top-middle), *C. ocellaris* from the Uaupés River (Upper Rio Negro) (top-right), *C. pleiozona* from the Jacundá River (lower Amazon River) (bottom-left), *C. orinocensis* from the Amazon estuary (bottom-middle), and *C. kelberi* from the Amazon estuary (bottom-right). *Photos: K. Winemiller.*

Cichla ocellaris seems to tolerate a greater range of water quality than other peacock bass, which may account for its large geographic distribution as compared to other species. For example, *C. ocellaris* is common in the acidic black waters throughout the Negro Basin in Brazil, Colombia, and Venezuela. In the Branco River, a clearwater tributary of the Negro in Brazil's Roraima state, this species inhabits the same kinds of river and floodplain habitats, but the water is only slightly acidic or has neutral pH.

Cichla ocellaris is widely distributed in side channels and floodplain lakes within the vast floodplain of the Amazon River. Water quality in these floodplain habitats ranges from acidic black water to near-neutral and slightly turbid water. In the Solimões River (Amazon River above the Negro confluence) and its tributaries, *C. ocellaris* is the only peacock bass species found in secondary channels and floodplain lakes. It also is the only peacock bass native to the Upper Madeira Basin (region above the San Antonio Falls). In the middle and lower Amazon, *C. ocellaris* often co-occurs with other *Cichla* species, especially *C. temensis* in the Rio Negro and certain habitats in the lower Madeira, and *C. pinima* in the floodplain of the lower Amazon to its estuary as well as lower reaches of tributaries. In the Araguaia-Tocantins Basin, *C. ocellaris* usually is found together with *C. piquiti*.

The only place where four species of peacock bass can be found together is the upper reaches of the Casiquiare and Upper Orinoco Basin, including tributaries such as the Mavaca and Ocamo rivers. In this region, *C. ocellaris* is most often found in floodplain lakes, sometimes together with *C. temensis* if the lake is deep enough. *Cichla ocellaris* and *C. orinocensis* are rarely found together in the same habitat, and may compete in shallow, nearshore habitats. The fourth species in this region, *C. intermedia*, essentially is restricted to river channels in habitats with flowing water. The distribution of *C. ocellaris* in the Orinoco River does not extend downstream much beyond the location where the Casiquiare has its origin, and the reason why this species has not colonized seemingly suitable habitats throughout the middle and lower Orinoco Basin is unclear. It may be the case that it is unable to compete with *C. orinocensis* in those habitats; however, we note that both of these species are broadly distributed in the Rio Negro Basin. It is rather puzzling that *C. ocellaris* is distributed throughout the Branco River all the way to its headwaters in the Rupununi Savanna District, throughout the Essequibo River Basin in Guyana, and within other coastal rivers of the Guiana Shield, but *C. orinocensis* is restricted to the lowermost reaches of the Branco.

Compared to the larger peacock bass species (*C. temensis*, *C. pinima*, and *C. piquiti*), *Cichla ocellaris* tends to occupy shallower water close to shore where there are submerged roots, branches, logs, or aquatic vegetation. This pattern of habitat segregation also is observed in rivers and lakes of the Upper Paraná Basin where both *C. ocellaris* and *C. piquiti* have been introduced. Among peacock bass species, *C. ocellaris* seems to have the greatest tolerance for turbidity, but there is a limit to this tolerance. Like all peacock bass species, transparent water is required for locating and tracking prey using vision. In muddy rivers, such as the Solimões-Amazon and lower Orinoco, peacock bass and other visually oriented predators are largely replaced by catfishes that can locate prey using olfaction (taste buds located on their barbels, lips, and skin) and electroreception (low-amplitude electroreceptors on their heads).

Cichla ocellaris var. *kelberi* taken from a reservoir on Paranapanema River, Upper Paraná Basin, Brazil. The species has been widely introduced throughout this basin. Note that this specimen lacks yellow spots on pelvic and anal fins but has them on the dorsal and caudal fins. *Photo: K. Winemiller.*

Introduced stocks

Given their tendency to thrive in constructed lakes, relatively wide tolerance of water quality variation (pH 4–7, depending on the regional stock) and popularity with anglers, *C. ocellaris* has been introduced into canals, ponds, and reservoirs in regions well beyond its natural distribution. *Cichla ocellaris* var. *ocellaris* now thrives in coastal canals in Georgetown, Guyana, and Miami, Florida [8–10]. *Cichla ocellaris* var. *kelberi* was extensively introduced and now supports important sport fisheries in reservoirs of the Upper Paraná and Paraguay basins [11, 12]. The many reservoirs in these basins have aided dispersal. This species also has been established in reservoirs of the São Francisco and other rivers of the Atlantic Forest region of southeastern Brazil [13, 14]. *Cichla ocellaris* supports a major sport fishery in Panama's Lake Gatun, a reservoir constructed on the Chagres River for the Panama Canal [15, 16]. We also received a recent report with photographic documentation from Dr. Carlos Lasso Alcala of the Instituto de Investigación de Recursos Biológicos Alexander von Humboldt in Colombia, that *C. ocellaris* has been introduced into reservoirs and lakes within the Magdalena River Basin in Colombia.

Population structure and abundance

In the upper Orinoco-Casiquiare region of Venezuela, a sample of *C. ocellaris* var. *monoculus* ranged in size from 19.2 to 34.6 cm and 0.2 to 0.8 kg (average size was 27.4 cm and 0.5 kg) [17]. In the Paraguá River of Bolivia, standard length of male *C. ocellaris* var. *pleiozona* ranged from 19 to 55 cm (average was 39 cm) and weight ranged from 0.07 to 2.65 kg (average was 1 kg) and females measured from 15 to 46 cm (average was 35 cm) and weighed 0.04 to 1.38 kg (average was 0.67 kg) [18]. The sex ratio was 55% males and 45% females.

The density of *C. ocellaris* var. *pleiozona* in the main channel of the Paraguá River was estimated at 1–44 fish per km, in side channels density was 5–53 fish per km, and in floodplain lakes (oxbows) it was 17–50 fish per km.

Studies of *C. ocellaris* var. *kelberi* in reservoirs of southern Brazil report that males and females attain similar body lengths [19, 20], and others found that males were generally larger [21]. In Lobo Reservoir, males averaged 30.5 cm and females 20.5 cm, and the length-weight relationship did not vary between the sexes [20]. In Panama's Lake Gatun, males and females have similar growth rates until reaching the first year, after which males grew faster than females, perhaps because males invest less energy in gonad maturation [22].

Feeding

All peacock bass feed predominately on fish and occasionally consume shrimp and other aquatic invertebrates. Developmental stages and seasonal availability of prey influence diet. Larvae begin feeding on zooplankton, such as rotifers and aquatic microcrustaceans, including daphnia, copepods, and ostracods. Small juveniles feed on aquatic insects and small shrimp, and as they grow, fish are consumed increasingly. Subadult and adult *C. ocellaris* of all varieties feed on fish almost exclusively.

Male *C. ocellaris* var. *monoculus* taken from a floodplain lake in the lower Branco River, Brazil. *Photo: K. Winemiller.*

Seasonal rainfall induces flood pulses in tropical rivers, and changes in water levels and flooding greatly influence predator-prey dynamics. In rivers that have a single, prolonged flood pulse each year, prey fish reproduce during the high-water phase and build up their populations within expansive aquatic habitats of floodplains. As water levels gradually fall and floodplains drain, fish densities increase within the reduced volume of aquatic habitat, and predator-prey interactions intensify. The early stages of the falling-water phase are when peacock bass in rivers exploit abundant prey and generally display an increase in body condition (i.e., become fatter) [23]. Peacock bass exploit the most profitable prey, which generally are the most abundant fish that are of the largest size capable of being ingested. As water levels continue to fall during the dry season, the size of prey in peacock bass stomachs decreases, presumably because abundance of larger and more profitable prey has been reduced by predation [24].

Examination of stomachs from 106 *C. ocellaris* from the Rupununi Savanna District of Guyana found 48 that were empty, 12 contained characiform fish, 7 had cichlids, 4 had shrimp, and 1 had catfish (*Corydoras* species) [8]. Nematodes were present in the guts of many fish, including juveniles as small as 4 cm.

Several studies have investigated diets of *C. ocellaris* var. *kelberi* from reservoirs in southern Brazil. *Cichla ocellaris* var. *kelberi* in Serra da Mesa reservoir consumed mostly characids (e.g., *Astyanax*, *Bryconops*, *Ctenobrycon*, and *Knodus* species) and cichlids (*Satanoperca jurupari* and juvenile *C. ocellaris*) [25]. In Leme Reservoir, *C. ocellaris* var. *kelberi* consumed fish, including cichlids (*Geophagus brasiliensis*) and a callicthyide catfish (*Hoplosternum littorale*) [26]. Cichlids dominated the diet of *C. ocellaris* var. *kelberi* in Lajes Reservoir, with over half the diet consisted of juvenile *C. ocellaris* var. *kelberi*, followed by tilapia (*Coptodon rendalli*), tetras (e.g., *Astyanax* species), and catfish (e.g., *Pimelodella eigenmanni*, *Rhamdia quelen*) [14].

Cannibalism

Given their predatory nature and fairly nondiscriminating diet, peacock bass are prone to cannibalism [21, 24, 27]. Early reports based on observations of *C. ocellaris* in Guyana, Hawaii, and Panama claimed that butterfly peacock bass are not cannibalistic [8, 28, 29]; however, these claims were based on examination of few specimens. One report even proposed that a function of the caudal ocellus and spots on the dorsal fin and body of young peacock bass is for species recognition so that parents may avoid eating their own offspring [29]. Obviously, it is not good for the fitness of parents to eat their offspring after they invested time and energy into gamete production and brood guarding. But there is a counterargument to this theory. For a predatory fish that relies on speedy attacks, any delay in attacking potential prey caused by assessment of a species recognition marking would increase the chance of a failed attempt, which clearly would reduce individual fitness.

The degree of cannibalism appears to be influenced by the availability of alternative prey. In Lajes Reservoir, juvenile *C. ocellaris* were a dominant component of adult diets, and this finding was attributed to overpopulation of *Cichla* and low prey availability [30]. Cannibalism was confirmed for *C. ocellaris* in Serra da Mesa Reservoir, especially during the operation phase of the hydroelectric dam when prey populations had declined following reservoir filling [25]. *Cichla ocellaris* showed high rates of cannibalism in Corumbá and Leme reservoirs (Brazil), two lakes lacking aquatic vegetation that typically provides refuge for small fish [31, 32]. In Leme, evidence of cannibalism increased from December to April, the period when peacock bass spawned [31, 33].

Effects on food webs

Because they are voracious predators with broad diets, peacock bass are capable of regulating populations of prey under certain circumstances. A famous example of this comes from a study conducted on Lake Gatun, a reservoir constructed on the Chagres River of the Panama

Canal [15]. In less than 5 years after the introduction of this exotic predator, half of the 14 native fish that had populated the lake were eliminated, three of the remaining species had reduced abundance, and only one large native cichlid (*Vieja maculicauda*) increased in abundance. Most of the affected fish were tetras (Characidae) and livebearers (Poeciliidae). Changes to the fish community resulted in a cascade of effects on the aquatic community. The elimination of small fish that consume aquatic invertebrates resulted in an increase in the density of mosquitos and higher incidence of the deadly *Plasmodium falciparum* form of malaria in people living near the lake. The reduced abundance of small fishes near the shoreline also resulted in lower abundance of fish-eating birds, such as kingfishers and herons. The peacock bass also reduced the abundance of *Melaniris chagresi* (Atherinidae), a pelagic silverside that feeds on zooplankton. This in turn led to changes in the zooplankton community and sharp reductions in species that feed on silversides, including tarpon (*Megalops atlanticus*) and black tern (*Chlidonias niger*). A study conducted in Lake Gatun 45 years after the introduction of peacock bass determined that *C. ocellaris* remains the most abundant predator in the lake and the collapse of the prey community has been sustained, with average abundance of native fish inhabiting shallow shoreline areas reduced by 96% compared to what it was before the introduction [16].

The cichlid, *Vieja maculicauda*, is native to Central America. This relatively large species was one of the only fish stocks not severely impacted by *C. ocellaris* after it was introduced into Lake Gatun, Panama. *Photo: K. Winemiller.*

Growth

Growth of butterfly peacock bass in the Rupununi Savanna region of Guyana was reported to be fastest during the rainy season, and slow during the dry season when food becomes scarce. In Guyana, juvenile *C. ocellaris* grew about 2.7 cm per month during 4 months of the rainy period [34]. Juveniles in a Hawaiian reservoir grew 1.5 cm per month, whereas juveniles from the same stock that were reared in aquaria grew 2.5 cm per month [35].

Due to many misidentifications of this species, the growth potential of *C. ocellaris* is uncertain. The International Gamefish Association (IGFA) all-tackle world record for *Cichla ocellaris* is recorded as 5.7 kg (12 lbs, 9 oz), but this record should be considered invalid because the fish was caught from the Chiguao River in Venezuela's Bolivar state. This river flows into Guri Reservoir which was constructed on the Caroní River, a blackwater tributary of the lower Orinoco River. Since *C. ocellaris* does not naturally occur in this region, and there are no records of stocking this species into Guri, it appears to be a misidentification of what most likely was a hybrid between *C. orinocensis* and *C. temensis*, the two species that were stocked into the reservoir and which are known to hybridize there (see Chapter 11).

C. ocellaris caught from a backwater of the Rewa River, Guyana. Most of these fish were probably nesting. *Photo: K. Winemiller.*

The next largest *C. ocellaris* recorded by the IGFA was caught in 2010 from the South River, Suriname, and weighed 5.67kg (12lbs, 8oz). Maps do not show a "South River" in Suriname, and because *C. temensis* was introduced into the Brokopondo Reservoir on the Suriname River, we cannot rule out the possibility that this fish could have been a hybrid. A fish caught in Cool Springs, Florida, weighed 4.54kg (10lbs). Since *C. ocellaris* appears to be the only species of peacock bass with self-sustaining stocks in the canals of south Florida, and all photos of Florida fish posted on the internet appear to be *C. ocellaris*, this likely is a valid record. However, it should be noted that *C. temensis* was introduced into canals of the region during the 1970s [36] but did not survive for long. It is possible that some hybridization could have occurred before *C. temensis* perished from the system.

Cichla ocellaris from the Brokopondo Reservoir, Suriname. *Photo: H. López-Fernández.*

The IGFA record for *C. kelberi* (here, this is considered a variety of *C. ocellaris*) is listed at 1.8kg (4lbs) for a fish from Paraibuna Dam, Brazil, and the IGFA record for *C. pleiozona* is listed at 4.58kg (10lbs) for a fish caught from Gatun Lake, Panama (the latter should be considered *C. ocellaris* var. *monoculus*). The largest *C. ocellaris* reported from Guyana was 50cm and 3.5kg (8lbs) [8], and the largest *C. ocellaris* var. *monoculus* reported from Peru was 50.4cm [37]. Again, these reports may be questionable given the potential for species misidentification.

Reproduction

Based on research on *C. ocellaris* in Guyana, both sexes were reported to mature at about 20cm and 0.2kg [8]. *Cichla ocellaris* reared in ponds in Brazil matured after 11–12months [38]. In Bolivian rivers, *C. ocellaris* var. *pleiozona* was reported to mature at about 17cm (males) and 17–20cm (females) [39]. In Brazilian reservoirs, *C. ocellaris* var. *kelberi* mature at about at 21cm [20]. Captive *C. ocellaris* in Hawaii reached sexual maturity at 1 year and about 30cm [35].

Butterfly peacock bass reproduction is fairly synchronized in rivers where there is strong wet/dry seasonality and less synchronized in lakes and reservoirs where environmental

A small male *C. ocellaris* var. *monoculus* from the Rio Negro, Brazil, with a nuchal hump and breeding coloration. *Photo: K. Winemiller.*

conditions vary less. In rivers and floodplain lakes, spawning is usually initiated near the end of the annual flood pulse. In the Rupununi Savanna District of Guyana, the wet season is from May through September, and gonads of most *C. ocellaris* were mature in April, but some fish had ripe gonads as early as December [8]. *Cichla ocellaris* var. *pleiozona* in Bolivia's Iténez and Mamoré rivers had mature gonads during the dry season (September–February), with a peak during November–January [40]. In Bolivia's Paraguá River, fish in floodplain lakes spawned about 1 month earlier than those in the river channel [20].

Spawning in artificial ponds and reservoirs can occur during any time of year, but often with one or more peaks in activity. In Panama's Lake Gatun, there are two peaks, one during the dry season (January–April) and one after the start of the rainy season (June–August). *Cichla ocellaris* var. *kelberi* from Três Marias, Campo Alegre, and Lobo reservoirs in southern Brazil reproduce from about November to April, with peaks coinciding with lower water temperatures [13, 19, 20]. In a Hawaiian reservoir, *C. ocellaris* was observed to have a relatively short period of peak spawning [28].

Prior to courtship and spawning, both sexes develop a more vivid, orange-red coloration in the region of the throat, chest, pelvic fins, anal fin, and lower half of the caudal fin [40].

Males develop a nuchal hump, a fatty deposit that bulges on the forehead. The nuchal hump probably advertises to females not only reproductive preparedness, but also the male's nutritional status (i.e., fitness), given that a history of successful feeding is required to obtain surplus energy for conversion to fat stores. This fatty deposit also may provide energy reserves that help sustain the male during the period of spawning and brood guarding when feeding ceases.

Female (top) and male (bottom) *C. ocellaris* caught while nesting in the Rewa River, Guyana. These fish show the bright red iris of breeding fish, and the male has a distinct nuchal hump. *Photo: K. Winemiller.*

Mature eggs (vitellogenic oocytes) of *C. ocellaris* var. *kelberi* are elliptical, measuring approximately 1.2×0.7mm [39]. The maximum diameter of mature eggs of *C. ocellaris* var. *monoculus* from the middle Solimões River was estimated to vary from 1.2 to 3.0mm [41]. Batch fecundity (number of ripe eggs per batch spawned) of *C. ocellaris* var. *monoculus* from the Solimões averaged 8624 [41]. Batch fecundity of *C. ocellaris* from rivers in Guyana ranged from 6000 to 10,000 and averaged about 8000 eggs [8]. *Cichla ocellaris* var. *pleiozona* in Bolivia had batch fecundity that ranged from 3712 to 10,355 for females weighing 0.5 and 1.4kg, respectively [18], and there was evidence that females could spawn more than once during a breeding season. Fish from Três Marias Reservoir in southeastern Brazil had batch fecundity ranging from 4450 for a female measuring 31.5cm total length, to 13,900 for a female measuring 43.5cm total length [39]. Average batch fecundity of *Cichla ocellaris* var. *kelberi* from Volta Grande Reservoir was estimated at 13,769 eggs, fish from Lobo Reservoir produced an average of 6072 eggs, and fish from Campo Alegre Reservoir averaged only 3100 eggs [19, 20, 42]. The estimated average batch fecundity for *C. ocellaris* in Lake Gatun was reported as 10,000 [22], and the estimated average for *C. ocellaris* in a Hawaiian reservoir was 5435 eggs [29].

Like all peacock bass, *Cichla ocellaris* is a substrate nester with biparental care of eggs and fry. Eggs are laid in rows on a solid substrate, usually a log or rock, and the male follows releasing milt. Rows of eggs are laid during repeated intervals that may last as long as 2.5 hours. The eggs hatch after 70–90 hours, and fry have an adhesive organ on the head that causes them to stick to surfaces. The parents prepare one or more nests in sand or clay using their chin and pectoral fins to clear debris and sediment. *Cichla ocellaris* var. *pleiozona* in the Paraguá River, Bolivia, were seen digging nests beneath branches at depths less than 1 m [18]. The nests are 0.2–1.5 m in diameter and 1.5–6 cm deep and may be located 2 m from the original egg deposition site in shallow water (0.2–0.3 m depth) [8], but in reservoirs the nest may be placed at depths up to 6 m [22]. The parents carry the fry to the nest in their mouths, and the adhesive organ sticks the fry to the nest substrate. Fry produce swimming motions and the brood appears as a wriggling mass. Their swimming motion, aided by the fanning of the nest by the parents using their pectoral fins, circulates water above the fry and thereby disperses dissolved waste products while bringing in water with more dissolved oxygen.

Cichla ocellaris eggs in nest on rocks in shallow water in a backwater of Rewa River, Guyana. *Photo: K. Winemiller.*

Location of a *C. ocellaris* nest in the Rewa River, Guyana. *Photo: K. Winemiller.*

The brood is defended aggressively by parents displaying vivid coloration that serves as a warning to would-be intruders. The fry may be moved several times from one nest pit to another. Aggregations of peacock bass nests have been observed in several tropical rivers and reservoirs [27]. Nest aggregations were reported for *C. ocellaris* var. *pleiozona* in the Paraguá River with approximately 18 nests contained within an area of 50 m^2 with as little as 1 m between nests of adjacent pairs [18]. These aggregations may serve as an added protective measure against egg and fry predation. After 4–6 days, the fry become free swimming and begin to feed on zooplankton while being guarded by the parents [8]. *Cichla ocellaris* have been observed guarding their broods for periods up to 10 weeks when the juveniles have grown to about 6–7 cm [22]. The parents feed very little while brood guarding, which results in a decline in fat reserves and body condition. About 80% of stomachs of fish from the Paraguá River were empty during the breeding season [18]. Nesting peacock bass will attack lures and baitfish with aggression, and these attacks apparently are motivated by instinct for brood defense rather than hunger.

Cichla ocellaris var. *monoculus* for sale in the fish market in Santarem, a city located at the mouth of the Tapajos River on the south bank of the lower Amazon River. *Photo: K. Winemiller.*

References

[1] S.C. Willis, J. Macrander, I.P. Farias, G. Orti, Simultaneous delimitation of species and quantification of interspecific hybridization in Amazonian peacock cichlids (genus *Cichla*) using multi-locus data, BMC Evol. Biol. 12 (2012) 96, https://doi.org/10.1186/1471-2148-12-96.

[2] S.C. Willis, I.P. Farias, G. Orti, Multi-locus species tree for the Amazonian peacock basses (Cichlidae: *Cichla*): emergent phylogenetic signal despite limited nuclear variation, Mol. Phylogenet. Evol. 69 (3) (2013) 479–490.

[3] S.C. Willis, One species or four? Yes! Or, arbitrary assignment of lineages to species obscures the diversification processes of Neotropical fishes, PLoS One 12 (2) (2017), e0172349.

[4] S.O. Kullander, H. Nijssen, The Cichlids of Suriname, Brill, Leiden, Netherlands, 1989.

[5] C.S.A.F. Camargo, M. Petrere Jr., Análise de risco aplicada ao manejo precaucionário das pescarias artesanais na região do reservatório da UHE-Tucuruí (Pará, Brasil), Acta Amazon. 34 (2004) 473–485.

[6] T.A. Barbosa, N.L. Benone, T.O. Begot, A. Goncalves, L. Sousa, T. Giarrizzo, L. Juen, L. Montag, Effect of waterfalls and the flood pulse on the structure of fish assemblages of the middle Xingu River in the eastern Amazon basin, Braz. J. Biol. 75 (2015) 78–94.

[7] D.B. Fitzgerald, M.H. Sabaj-Perez, L.M. Sousa, A.P. Gonçalves, L.R. Py-Daniel, N.K. Lujan, J. Zuanon, K.O. Winemiller, J.G. Lundberg, Diversity and community composition of rapids-dwelling fishes of the Xingu River: implications for conservation amid large-scale hydroelectric development, Biol. Conserv. 222 (2018) 104–112.

[8] R.H. Lowe-McConnell, The cichlid fishes of Guyana, South America, with notes on their ecology and breeding behavior, Zool. J. Linn. Soc. 48 (1969) 255–302.
[9] P.L. Shafland, An overview of Florida's introduced butterfly peacock bass (*Cichla ocellaris*) sportfishery, Natura 96 (1993) 26–29.
[10] P.L. Shafland, The introduced butterfly peacock (*Cichla ocellaris*) in Florida. II. Food and reproductive biology, Rev. Fish. Sci. 7 (1999) 95–113.
[11] A.A. Agostinho, S.M. Thomaz, L.C. Gomes, Conservation of the biodiversity of Brazil's inland waters, Conserv. Biol. 19 (3) (2005) 646–652.
[12] A.A. Agostinho, F.M. Pelicice, A.C. Petry, L.C. Gomes, H.F. Julio Junior, Fish diversity in the upper Paraná River basin: habitats, fisheries, management and conservation, Aquat. Ecosyst. Health Manage. 10 (2) (2007) 174–186.
[13] A.L.B. Magalhães, Y. Sato, E. Rizzo, R.M.A. Ferreira, N. Bazzoli, Ciclo reprodutivo do tucunaré *Cichla ocellaris* (Schneider, 1801) na represa de Três Marias, MG, Arq. Bras. Med. Vet. Zootec. 48 (1) (1996) 85–92.
[14] G. Neves dos Santos, A. Filippo, L. Neves dos Santos, F. Gerson Araújo, Digestive tract morphology of the Neotropical piscivorous fish *Cichla kelberi* (Perciformes: Cichlidae) introduced into an oligotrophic Brazilian reservoir, Rev. Biol. Trop. 59 (3) (2011) 1245–1255.
[15] T.M. Zaret, R.T. Paine, Species introduction in a tropical lake, Science 182 (1973) 449–455.
[16] D.M.T. Sharpe, L.F. De Leon, R. Gonzalez, M.E. Torchin, Tropical fish community does not recover 45 years after predator introduction, Ecology 98 (2017) 412–424.
[17] D.B. Jepsen, K.O. Winemiller, D.C. Taphorn, D. Rodriguez Olarte, Age structure and growth of peacock cichlids from rivers and reservoirs of Venezuela, J. Fish Biol. 55 (1999) 433–450.
[18] H. Muñoz, P. Damme, F. Duponchelle, Breeding behaviour and distribution of the tucunare *Cichla* aff. *monoculus* in a clear water river of the Bolivian Amazon, J. Fish Biol. 69 (2006) 1018–1030.
[19] S. Chellappa, M.R. Camara, N.T. Chellappa, M.C.M. Beveridge, F.A. Huntingford, Reproductive ecology of a neotropical cichlid fish, *Cichla monoculus* (Osteichthyes: Cichlidae), Braz. J. Biol. 63 (2003) 17–26.
[20] J.E. Souza, E.N. Fragoso-Moura, N. Fenerich-Verani, O. Rocha, J.R. Verani, Population structure and reproductive biology of *Cichla kelberi* (Perciformes, Cichlidae) in Lobo Reservoir, Brazil, Neotrop. Ichthyol. 6 (2) (2008) 201–210.
[21] L.M. Gomiero, F.M.S. Braga, Relação peso-comprimento e fator de condição para *Cichla* cf. *ocellaris* e *Cichla monoculus* (Perciformes, Cichlidae) no reservatório de Volta Grande, Rio Grande, Acta Sci. Biol. Sci. 25 (1) (2003) 79–86.
[22] T.M. Zaret, Life history and growth relationships of *Cichla ocellaris*, a predatory South American cichlid, Biotropica 12 (1980) 144–157.
[23] D.J. Hoeinghaus, K.O. Winemiller, C.A. Layman, D.A. Arrington, D.B. Jepsen, Effects of seasonality and migratory prey on body condition of *Cichla* species in a tropical floodplain river, Ecol. Freshw. Fish 15 (4) (2006) 398–407.
[24] D.B. Jepsen, K.O. Winemiller, D.C. Taphorn, Temporal patterns of resource partitioning among *Cichla* species in a Venezuelan blackwater river, J. Fish Biol. 51 (1997) 1085–1108.
[25] J.L.C. Novaes, E.P. Caramaschi, K.O. Winemiller, Feeding of *Cichla monoculus* Spix, 1829 (Teleostei: Cichlidae) during and after reservoir formation in the Tocantins River, Central Brazil, Acta Limnol. Bras. 16 (1) (2004) 41–49.
[26] G.A. Villares Junior, L. Gomiero, Feeding dynamics of *Cichla kelberi* Kullander & Ferreira, 2006 introduced into an artificial lake in southeastern Brazil, Neotrop. Ichthyol. 8 (2010) 819–824.
[27] K.O. Winemiller, Ecology of peacock cichlids (*Cichla* spp.) in Venezuela, J. Aquaric. Aquat. Sci. 9 (2001) 93–112.
[28] W. Devick, Life history of the Tucunare, *Cichla ocellaris*. State of Hawaii Federal Aid Project F-4R-17, 1971.
[29] T.M. Zaret, Inhibition of cannibalism in *Cichla ocellaris* and hypothesis of predator mimicry among South American fishes, Evolution 31 (1977) 421–437.
[30] L.N. Santos, A.F. Gonzáles, F.G. Araújo, Dieta do tucunaré-amarelo *Cichla monoculus* (Bloch & Schneider) (Osteichthyes, Cichlidae), no reservatório de Lajes, Rio de Janeiro, Brasil, Rev. Bras. Zool. 18 (2001) 191–204.
[31] G.A. Villares Junior, L.M. Gomiero, R. Goitein, Biological aspects of *Schizodon nasutus* Kner, 1858 (Characiformes, Anostomidae) in the low Sorocaba river basin, São Paulo state, Brazil, Braz. J. Biol. 71 (3) (2011) 763–770.
[32] R. Fugi, K.G. Luz-Agostinho, A.A. Agostinho, Trophic interaction between an introduced (peacock bass) and a native (dogfish) piscivorous fish in a neotropical impounded river, Hydrobiologia 607 (2008) 143–150.
[33] L.M. Gomiero, G.A. Villares Junior, F. Naous, Reproduction of *Cichla kelberi* Kullander and Ferreira, 2006 introduced into an artificial lake in southeastern Brazil, Braz. J. Biol. 69 (2009) 175–183.

[34] R.H. Lowe-McConnell, The fishes of the Rupununi savanna district of British Guiana, South America, part 1. Ecological groupings of fish species and effects of the seasonal cycle on the fish, J. Linn. Soc. Zool. 45 (1964) 103–144.
[35] W.S. Devick, Life history study of the tucunaré *Cichla ocellaris*, in: Federal Aid in Sportfish Restoration Project F-9-1, Job Completion Report, Hawaii Department of Land and Natural Resources, Honolulu, 1972.
[36] P.L. Shafland, B.D. Hilton, Introduction of peacock bass (*Cichla* spp.) in southeast Florida canals. First annual performance report, Florida Game and Fresh Water Fish Commission, Tallahassee, 1985.
[37] P. Cala, E. Gonzalez, M.P. Varona, Aspectos biologicos y taxonomicos del tucunare´, *Cichla monoculus* (Pisces: Cichlidae), Dahlia 1 (1996) 23–37.
[38] O. Fontenele, Contribução para o conhecimiento de biologia dos tucanarés, Actinopterygii Cichlidae, em cativiero. Aprelhlo de reproducao. Habitos de desova incubaçã, Rev. Bras. Biol. 10 (1950) 503–519.
[39] F.T. Normando, F.P. Arantes, R.K. Luz, R. Thome, E. Risso, Y. Sato, N. Bazzoli, Reproduction and fecundity of tucunaré, *Cichla kelberi* (Perciformes: Cichlidae), an exotic species in Três Marias Reservoir, Southeastern Brazil, J. Appl. Ichthyol. 25 (2009) 299–305.
[40] F. Duponchelle, F. Carvajal, J. Núñez, J.F. Renno, Historia de vida del tucunaré, *Cichla cf monoculus*, en la Amazonía boliviana, in: Primer coloquio de la red de investigation sobre la ictiofauna Amazonica (RIA), Iquitos, Peru, 2005.
[41] R. Lira de Souza, D.L. Silva, A. Prado-Valladares, R. Rocha, M. Ferreira, H. Queiroz, Gonodal development of the peacock bass *Cichla monoculus* (Perciformes: Cichlidae) in the Middle Solimões, Sci. Mag. UAKARI 7 (2011) 41–55.
[42] L. Gomiero, F.M.dS. Braga, Feeding of introduced species of *Cichla* (Perciformes, Cichlidae) in Volta Grande reservoir, River Grande (MG/SP), Braz. J. Biol. 64 (2004) 787–795.

CHAPTER

3

Orinoco butterfly peacock bass, *Cichla orinocensis* (Humboldt, in Humboldt & Valenciennes, 1821)

Cichla orinocensis

Origin of name

Latin "*orinocensis*" meaning "originating in Orinoco"

Synonyms

Cichla argus, Humbolt & Valenciennes, 1821

Common names

Orinoco butterfly peacock bass, Three-spot peacock bass, Tucunaré borboleto, Pavón mariposa, Pavón tres estrellas, Pavón amarillo, Marichapa

https://doi.org/10.1016/B978-0-323-85157-2.00008-1

Geographic distribution

Cichla orinocensis is found in clearwater and blackwater tributaries of the Orinoco River in Colombia and Venezuela as well as blackwaters of the Rio Negro and clear waters of the Branco River, a major tributary of the Negro that drains Roraima state in northeast Brazil.

Natural history

Inhabits rivers, creeks, and floodplain lakes in shallow areas without current near shorelines where there are rocks, logs, and submerged vegetation. Tolerates wide range of pH, conductivity, and turbidity, and is found in both clearwater and blackwater systems. Primarily piscivorous.

By March of 1984, I (KW) had been in Venezuela for 3 months conducting field research for my doctoral dissertation on the community ecology of tropical fishes. I would stay in Venezuela for a full year, hosted by Don Taphorn, a professor at the University of the Western Llanos in Guanare, the capitol city of Portuguesa state. At that time, Don, a leading Neotropical ichthyologist, was completing an inventory of fish diversity in the Llanos for his own doctoral dissertation at the University of Florida. The previous year, I had met Don at the annual conference of the American Society of Ichthyologists and Herpetologists, and he graciously invited me to stay in his home while I studied the ecology of fishes in the Llanos. Fortunately, I was able to secure a grant from the National Geographic Society to fund my project. My study sites in the Llanos contained remarkable fish diversity—piranhas, tetras, various bizarre catfish, weakly electric knifefish, and my favorite group, cichlids. Unfortunately, my sites didn't have any peacock bass. I was told that in order to find peacock bass, you had to go south to areas with slow-moving, clearwater rivers.

I had the opportunity to visit Las Majaguas, a reservoir constructed only a few years prior by impounding the Cojedes River and then diverting water via canals into a basin formed by earthen dams. Notably, the lake had been stocked with both *C. orinocensis* and *C. temensis*. At last, I had the chance to catch the legendary peacock bass! I could barely contain my enthusiasm as Don and two colleagues left me by the lake to embark on a daylong fish collecting expedition. I carried rod and reel, a 5-gal bucket, and a backpack with water, food and, of course, a few fishing lures. I hiked around the lake shoreline for about 6 km, casting in every cove along the way. I caught piranhas, pike cichlids (*Crenicichla lugubris*), a parrot cichlid (*Hoplarchus psittacus*), severums (*Heros severus*), and a beautiful oscar (*Astronotus ocellatus*) with glistening orange scales that I kept alive in the bucket to take back for one of Don's aquariums. I fished the entire day, and only caught a couple of relatively small butterfly peacock bass, which I released back to the lake. It was a fantastic day in solitude, catching fish that I read about during my youth as a budding tropical fish hobbyist. As the afternoon waned, I headed back to the road to meet Don, making a cast here and there as I retraced my route along the shoreline. As I was walking and gazing at the ground, my arm growing weary from carrying the bucket full of water, I caught a glimpse of splashing in the shallow water near a log at the water's edge. I looked up and could see tiny fish skittering across the surface in front of some ripples that obviously were made by a moving predator. I hastily put down the bucket and backpack, tied on a small yellow jig, crouched low, and crept toward the lakeshore. One cast, right on the spot, and—wham!—fish on! I could tell the fish was big—much larger than anything else I had caught that day. It made long runs in the shallow water, and I hoped the line and the fish wouldn't get ensnarled amid the logs, sticks, and mats of floating vegetation in the cove. Eventually, I gently eased the fish onto shore—a beautiful Orinoco butterfly peacock bass—5 kg (10 1/2 lbs)! When Don and the others met me, they couldn't believe their eyes. Don had never seen a *Cichla orinocensis* that big, and we decided to preserve it for deposition in his university's natural history collection where it would be available for future scientific studies.

I worked diligently on my project that year, because I couldn't be certain if I would ever have another opportunity to visit the tropics. As it turned out, I've had the privilege of conducting research in tropical regions across the globe throughout my career. A segment of my ecological research has involved peacock bass, and in pursuit of new knowledge related to *Cichla*, I've been able to catch all of the known species, plus one that is new to science (see Chapter 10). I returned to Las Majaguas a few times with my friend Aniello Barbarino, who in those days was a student at the university and later became a government fisheries biologist. Aniello's home town is San Rafael de Onoto, just a stone's throw from Las Majaguas Reservoir. We fished from a small aluminum boat with Aniello's friend Pedro, and we caught both Orinoco butterfly (pavón mariposa) and speckled (pavón lapa) peacock bass. The speckled peacocks were always small, and at first, I thought that it was the smaller of the two species. I would later discover how very wrong was that initial impression! Apparently, the Orinoco butterfly peacock bass is more tolerant of water with neutral pH and high conductivity than the speckled peacock, the latter generally inhabiting blackwater and clearwater rivers with pH lower than six. The speckled peacock bass grows large in the acidic water of Guri Reservoir, which was created when a hydroelectric dam was constructed on the blackwater Caroní River, which lies on the Guiana Shield in Venezuela's Bolivar state. Las Majaguas receives water from the Cojedes River, which flows from the Andes and consequently has a fairly neutral pH, moderate conductivity, and relatively high nutrient and sediment loads. Environmental conditions in Las Majaguas and other reservoirs in the region do not appear to be conducive to growth of the speckled peacock bass, but as I learned from my first experience fishing for peacock bass, the Orinoco butterfly peacock can thrive and even grow to sizes not normally observed in their native habitats.

KW with a large Orinoco butterfly peacock bass (*C. orinocensis*) caught from Las Majaguas Reservoir, Venezuela, in 1984. *Photo: K. Winemiller.*

Identification

Cichla orinocensis has coloration, body shape, and average size similar to *C. ocellaris*, and both of these species are often referred to as the butterfly peacock bass. The two species have frequently been confused and misidentified in the scientific literature as well as in images posted by anglers on the internet. Early ichthyological studies synonymized *C. orinocensis* with *C. ocellaris* (Regan 1906). Several early publications describing the natural history of peacock bass considered *C. orinocensis* to be variety of *C. ocellaris*, but distinct difference between these two species have since been well established by morphological, genetic, and ecological studies [1–5].

Cichla orinocensis from clear water of the Cinaruco River, Venezuela. *Photo: H. López-Fernández.*

Cichla orinocensis from the blackwater of the Atabapo River in southern Venezuela. *Photo: C. Montaña.*

Cichla orinocensis differs from *C. ocellaris* by the absence of the three vertical bars that are diagnostic for identification of *C. ocellaris*. Three large, sometimes irregularly shaped but generally round, black spots on the sides of the body also characterize *C. orinocensis*. Each of the three spots is located approximately at the body midline, with each outlined by a bright yellow or gold ring and sometimes having a scattering of smaller yellow or gold specks inside the ring. The caudal ocellus is similarly colored and placed along the midline in the tail fin. The common names butterfly peacock (pavón mariposa) and three-stars (pavón tres estrellas) originate from these conspicuous ringed spots. However, juveniles and subadult fish sometimes lack strong development of the three spots, and instead have three triangular-shaped vertical bars in the position of bars 1, 2, and 3 in a manner superficially similar to adult *C. ocellaris* and several other *Cichla* species. These vertical bars are often a source of confusion when discriminating between these two species, but the bars in *C. orinocensis* juveniles and subadults are always thicker in the center and narrow to a point as they extend a short distance above and below the midline. Oftentimes, these vertical bars show evidence of yellow or gold borders. Even some large adult *C. orinocensis* may show a very faint pattern of a vertical bar extending above and below their prominent ocellated spots. Another difference between *C. orinocensis* and *C. ocellaris* is the presence of a discontinuous lateral line in the

Cichla orinocensis from a creek in the Casiquiare River floodplain. This specimen shows vertical bars instead of ocellated spots and also has a profusion of cloudy yellow spots all over the head, body, dorsal fin, and upper half of the caudal fin. *Photo: S. Willis.*

former versus a lateral line that is usually, but not always continuous in the latter. In addition, *C. orinocensis* lacks the abdominal blotching that is a distinct characteristic of *C. ocellaris*.

The background coloration of the head and body of the Orinoco butterfly peacock bass ranges from dull gray to golden yellow to olive green to green, usually darker in the dorsal region and lighter yellow or white in the belly region. Some fish may have tiny black dots scattered all over the body. Some fish living in extreme clear water or aquaria show cloudy white or yellow spots or cloudy wormy patterns all over the head, body, and fins. In some blackwater systems, such as Guri Reservoir on the lower Caroní River in Venezuela, the background coloration may be bronze, with cloudy white or yellow spots scattered over the head, body, and fins. Similar to *C. ocellaris*, the Orinoco butterfly peacock bass lacks any black markings (except for tiny dots observed on some fish) on the head, including the gill cover. The eye iris is orange or red, turning to a brilliant crimson in breeding fish. The dorsal fin and top half of the caudal fin usually are gray or gray-blue with light blue spots and streaks. The pelvic, anal, and lower half of caudal fin are orange or red.

Specimen of *C. orinocensis* showing many tiny black spots scattered on the body. *Photo: C. Montaña.*

Juveniles less than 5 cm have three black spots on the sides that lack ocellation, and the third of these spots blends into a black lateral stripe that runs to the base of the caudal fin. This black stripe spanning the posterior half of the body is also observed in juvenile *C. ocellaris*. In several other *Cichla* species, this black stripe extends from the eye continuously to the base of the caudal fin. As juveniles grow, the three spots on the sides become larger and often extend vertically to form bars as described before. Juveniles up to about 6 cm may have a faint postorbital stripe, but this disappears as fish grow. As observed in juveniles of most *Cichla* species, the dorsal, caudal, and anal fins of juvenile *C. orinocensis* are transparent gray with clear spots.

Immature *C. orinocensis* from the Cinaruco River, Venezuela. *Photo: C. Montaña.*

Juvenile *C. orinocensis* in the authors' (KW, LKW) home aquarium.

Distribution and habitat

Cichla orinocensis is found in clearwater and blackwater tributaries of the Orinoco River in Colombia and Venezuela as well as blackwaters of the Rio Negro and clear waters of the Branco River, a major tributary of the Negro that drains Roraima state in northeast Brazil. There are genetic differences between *C. orinocensis* from the middle and lower Negro and those from the upper Negro and Orinoco Basin [2]. These fish are indistinguishable based on

Natural distribution of *Cichla orinocensis* in Brazil, Colombia, and Venezuela.

morphology and coloration, and genetic differences involving mitochondrial DNA markers, which are maternally inherited, seem to indicate a history of hybridization with *C. ocellaris* var. *monoculus* in the middle-lower Negro lineage. *Cichla orinocensis* appears to be absent historically from the easternmost tributaries of the Orinoco that drain the Guiana Shield, such as the Caroní River, and this is likely due to the presence of cataracts that historically blocked dispersal. The butterfly peacock bass was stocked in Lago Guri, a reservoir with extreme blackwater conditions constructed on the Caroní River. To the north, this species is found in streams of the eastern Llanos, including Morichal Largo and others that flow into the Orinoco Delta.

Following construction of a system of dikes and ponds in Venezuela's Apure state during the 1970s, the Orinoco butterfly peacock bass became established in creeks and lagoons in the region's extensive floodplains. Although this region of the Llanos is dominated by whitewater conditions (neutral pH, relatively high conductivity) associated with clay-rich alluvium transported from the Andes Mountains by the Apure and Arauca rivers, slow-moving water in creeks and standing water in ponds allow suspended sediments to settle out. The result is productive habitat with transparent water and abundant prey that support Orinoco butterfly peacock bass.

An Orinoco butterfly peacock bass swimming around submerged logs and sticks, habitat components favored by this species. *Photo: K. Winemiller.*

Orinoco butterfly peacock bass seem to tolerate a greater range of water conductivity, pH, and turbidity than most other *Cichla* species. *Cichla ocellaris* is the only other peacock bass species that can tolerate a comparable range in the quality of water. *Cichla orinocensis* was recently introduced into Venezuela's Lake Maracaibo, a natural lake that receives water from rivers draining the Andes that gradually transitions into an estuary on the Caribbean coast. The species apparently has become established there and supports a modest sport fishery. We received a recent report that *C. orinocensis* has been stocked in lakes and reservoirs within the Magdalena River Basin in Colombia, a region lacking any native peacock bass species. After *C. orinocensis* were stocked in Guri Reservoir in the 1970s, at least some of them hybridized with *C. temensis* [2]. These hybrids sometimes have a horizontal stripe and vertical bars that may cause them to be misidentified as *C. ocellaris* or *C. intermedia*.

Within the Orinoco Basin, *Cichla orinocensis* generally coexists with *C. temensis* and *C. intermedia*, but also may be found in smaller bodies of water, such as creeks and small floodplain pools, where those two species are absent. The three species coexist in the Aguaro, Capanaparo, Cinaruco, Casiquiare, Caura, Ventuari, and Upper Orinoco rivers in Venezuela

Flooded forest in the Rio Negro floodplain is excellent habitat for *C. orinocensis*.

A local subsistence fisherman's catch of Orinoco butterfly peacock bass (*C. orinocensis*), speckled peacock bass (*C. temensis*), and freshwater barracuda (*Acestrorhynchus falcirostris*) from the Pasiba River, a blackwater tributary of the Casiquiare in southern Venezuela. *Photo: C. Montaña.*

and the Bita, Tomo, and other rivers in the Colombian Llanos. The Orinoco butterfly peacock bass tends to occupy shallow habitats close to the shore in river channels and floodplain lakes. In the Cinaruco River, where peacock bass have been studied extensively [5–8], *C. orinocensis* and *C. temensis* are both common within the main river channel and floodplain lakes; however, *C. orinocensis* tends to be more abundant in lakes and often is found in creeks to the exclusion of *C. temensis*. The Orinoco butterfly peacock bass is almost always captured within or near submerged sticks, logs, and vegetation. This species generally avoids habitats with swiftly flowing water, whereas *C. temensis* is commonly captured in areas near current, especially near rocks or logs that hinder flow, and *C. intermedia* is almost always captured from channel habitats with moderate to swift flow and near rocks or logs. All size classes of the Orinoco butterfly peacock bass enter flooded forests and savannas during the wet season.

The Orinoco butterfly peacock bass also occurs in blackwater and clearwater habitats throughout the Rio Negro Basin in the Brazilian Amazon, areas where *C. temensis* and *C. ocellaris* var. *monoculus* also are found. In this region, *C. orinocensis* and *C. ocellaris* var. *monoculus* are commonly captured from shallow areas near shore where there is submerged structure, but oftentimes only one species is present at a given location. The reasons for this are unclear, because both species can be found in channels and floodplain lakes and tolerate a wide range of water qualities. For example, we have only observed *C. orinocensis* in two clearwater creeks (Agua Boa and Cambréua) while snorkeling during the course of a dozen trips to the lower Rio Negro and lower Branco. Our frequent visits to a lake in the floodplain of the lower Branco River have consistently yielded *C. ocellaris* var. *monoculus* but never *C. orinocensis*. In many rivers and lakes of the Negro-Branco Basin, *C. temensis* is captured alongside *C. orinocensis* or *C. ocellaris* var. *monoculus* but not both. Water quality does not seem to explain the occurrence of either species, because both can be found in clearwater or extreme blackwater conditions. Given their similar size and habitat affinities, it may be that *C. orinocensis* and *C. ocellaris* var. *monoculus* compete; however, it presently is unclear what conditions determine competitive outcomes.

The eye and ocellated spots of this Orinoco butterfly peacock bass seem to almost glow within the clear water of Agua Boa, a densely vegetated, clearwater creek in the floodplain of the Rio Negro. *Photo: K. Alofs.*

A nice Orinoco butterfly peacock bass caught from Agua Boa. *Photo: K. Winemiller.*

Other fish commonly caught in habitats where butterfly peacock bass are found include other types of cichlids (*Crenicichla lugubris, Hoplarchus psittacus*), pike characins (*Acestrorhynchus* spp., *Boulengerella* spp.), and piranhas (*Serrasalsmus gouldingi, S. rhombeus,* and *S. manueli*). In the Rio Negro Basin, the arowana (*Osteoglossum* spp.) is sometimes caught when fishing for peacock bass (arowana are not native to the Orinoco Basin). When snorkeling, it is interesting to see schools of beautiful and seemingly delicate tropical fishes, such as angelfish (*Pterophyllum scalare*) that are well known to aquarists, swimming in the same habitats occupied by voracious peacock bass.

The black piranha, *Serrasalmus rhombeus* (sometimes called the red-eye piranha), is common in most rivers where Orinoco butterfly peacock bass thrive. *Photo: K. Winemiller.*

Feeding

Orinoco butterfly peacock bass are voracious predators that are particularly adept at attacking small fish in shallow water near submerged brush along shorelines or within flooded forests and savannas. This species is encountered in small creeks much more frequently than other peacock bass species within its geographic distribution, and this probably is because of its natural ability to pursue prey in tight spaces. Orinoco butterfly peacock bass sometimes hunt in small packs, and several fish can be captured by repeatedly casting a lure into the same location. Orinoco butterfly peacock bass often flush small fish from structure near shore into deeper water where large speckled peacock bass (*C. temensis*) "sit and wait" for an opportunity to attack fleeing prey. It is our belief that the popular method of casting lures as close as possible to shoreline structure and then retrieving in a manner that produces erratic movements is highly effective for catching speckled peacock bass, because it simulates a frantic fish that has been flushed from cover by attacking Orinoco butterfly peacock bass. Large Orinoco butterfly peacock bass often will pursue prey into deeper waters away from shore, but smaller fish seem reluctant to do so, probably because of the threat of predation by large

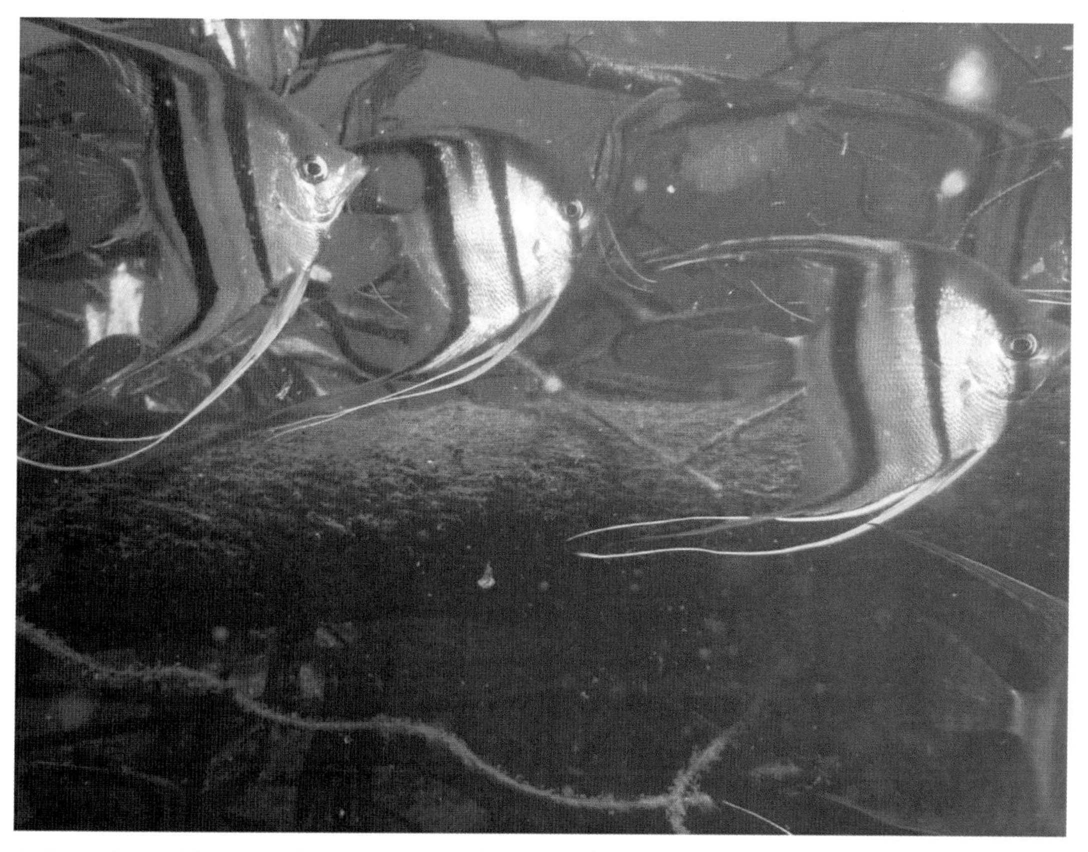

A school of angelfish, *Pterophyllum scalare*, in the clear water of a creek in the Rio Negro floodplain. Popular in the aquarium hobby, these seemingly delicate fish coexist with peacock bass and other predators by inhabiting shoreline areas with lots of submerged branches where they can deftly outmaneuver larger fish. *Photo: K. Winemiller.*

An Orinoco butterfly peacock bass whose "eyes were bigger than its stomach." These fish were found interlocked in a struggle on the bank of the Cinaruco River, Venezuela. The other fish is a pike characin, *Boulengerella cuvieri*. *Photo: D. Jepsen.*

speckled peacock bass lurking just offshore. We have experienced attacks by speckled peacock bass on Orinoco butterfly peacock bass that were hooked near shore and reeled toward deeper water near the boat.

The most detailed diet information for Orinoco butterfly peacock bass is from research conducted in the Venezuelan Llanos. *Cichla orinocensis* from the La Guardia River were found to feed heavily on fish, with at least 46 species from 17 families documented in stomach contents [9]. *Cichla orinocensis* from the Aguaro River (Llanos) consumed at least 20 different fish species, with cichlids and characiforms contributing the highest percentages [10]. Analysis of stomach contents of Orinoco butterfly peacock from the Cinaruco River in Venezuela revealed a diet dominated by fish, with insects contributing less than 5% of the diet by frequency of occurrence [7]. Nearly 50 fish species from 16 families were identified, with characiform fishes consumed most frequently. Depending on the season, from 60% to over 80% of stomachs

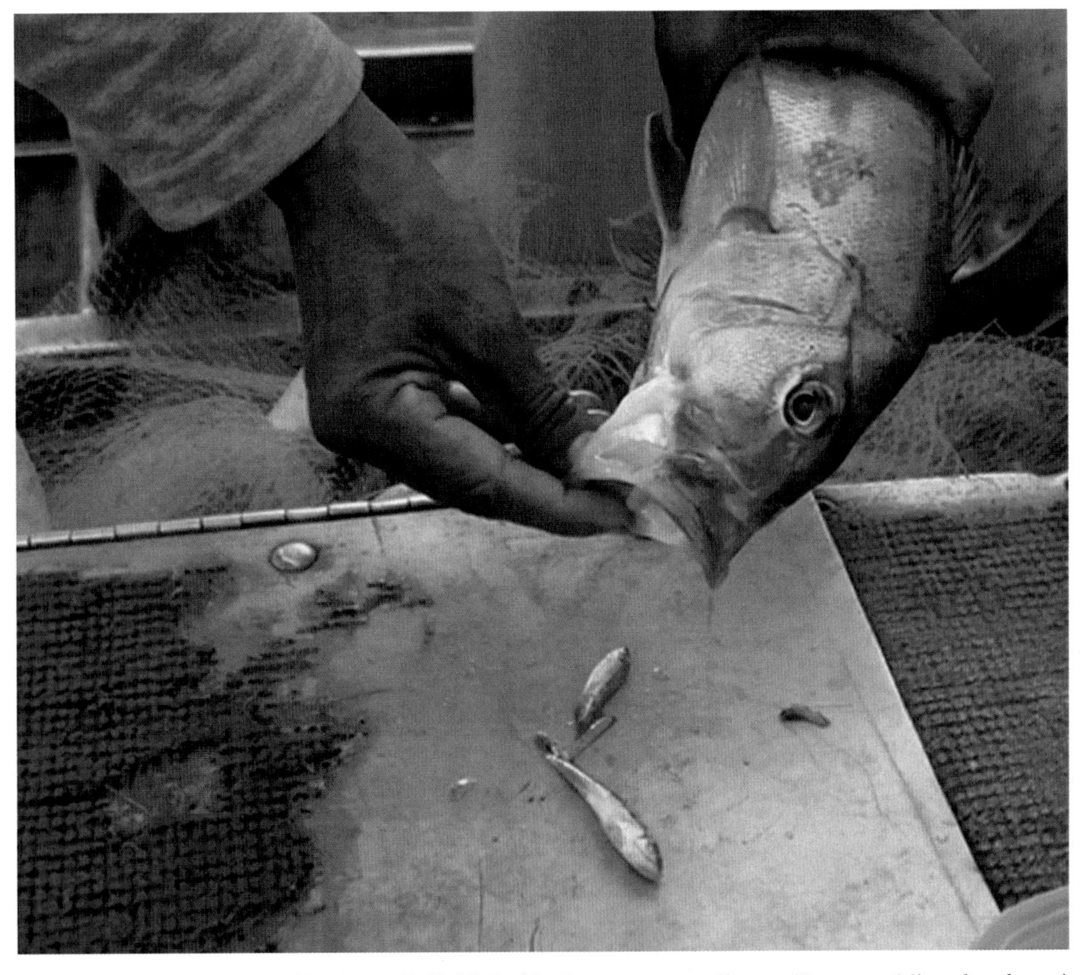

Examining stomach contents of *C. orinocensis*. Visible in this photo are two small tetras (*Bryconops*, *Microschemobrycon*) and a small peacock bass. *Photo: C. Montaña.*

were empty, which indicates relatively infrequent feeding bouts. Orinoco butterfly peacock bass consumed larger prey during the early portion of the falling-water period, but as water levels continued to fall during the dry season, the average size of consumed prey declined [7]. This trend presumably results from gradual depletion of preferred large prey as water levels drop and fish become more concentrated at higher densities, encounter rates between predators and prey increase, and predation mortality of prey increases. The same trend of declining prey size in diets of *C. orinocensis* was observed during research conducted in the La Guardia and Ventuari rivers [9].

Of course, larger peacock bass can consume larger prey. A positive relationship between predator length and prey length was reported for *C. orinocensis* from the Cinaruco River [7], Aguaro River [8], and Guri Reservoir [11]. Juveniles smaller than 5 cm feed mostly on zooplankton and aquatic insects, and fish become more important in the diet as the peacock bass grow [7].

Diet overlap between *C. orinocensis* and *C. temensis* in the Cinaruco was high during periods of falling and rising water, and overlap was relatively low during the dry season when water levels were lowest. In contrast, diet overlap between *C. orinocensis* and *C. intermedia* was zero during the falling-water period, negligible during the rising-water period, and high during the low-water period. *Cichla orinocensis* and *C. intermedia* are similar sizes, but *C. temensis* grows much larger and therefore consumes a greater range of prey sizes. *Cichla temensis* and *C. orinocensis* tend to occur together in river channels and floodplain lakes, whereas *C. intermedia* is largely restricted to channel habitats with flowing water and submerged structures.

Stomach contents were examined from samples of Orinoco butterfly peacock bass from two rivers in the Upper Orinoco Basin, a region of the Guiana Shield dominated by rainforest. Orinoco butterfly peacock bass from the Ventuari River consumed 59 prey categories, with 55 fish species from 20 different families comprising 83% of the diet by frequency of occurrence, and shrimp and aquatic insects making up the remainder [12]. Characids, hemiodontids, and cichlids were most frequent in the diet. Among fish examined from the Ventuari River, 75% had empty stomachs. In the Pasimoni River, an extreme blackwater tributary of the lower Casiquiare, the diet of Orinoco butterfly peacock bass consisted entirely of small characid fish (*Hemigrammus analis* and *H. vorderwinkleri*).

Orinoco butterfly peacock bass in Las Majaguas Reservoir consumed nearly equal proportions of fish and shrimp (*Macrobrachium amazonicum*) [13]. *Cichla orincensis* in Masparro Reservoir (Venezuela) fed on a cichlid (*Caquetaia kraussii*), characid (*Roeboides dientonito*), and shrimp, and these comprised over 50% of the diet [14]. Orinoco butterfly peacock bass in Guri Reservoir fed almost entirely on fish, with cichlids, characids, and hemiodontids dominant in the diet [11].

Cannibalism

Relatively low incidence of cannibalism by Orinoco butterfly peacock bass was documented by analysis of stomach contents of fish from the Cinaruco, La Guardia, and Ventuari rivers [7, 9]. This cannibalism may have been associated with fish confined to isolated water bodies in floodplains where prey became depleted [7]. Higher levels of cannibalism were reported for *C. orinocensis* in Guri and Masparro reservoirs, ecosystems that have much lower fish diversity and lower productivity than natural river-floodplain systems [14, 15].

Predation mortality

Freshwater dolphins (*Inia geoffrensis*), giant river otters (*Pteronura brasiliensis*), and spectacled caiman (*Caiman crocodilus*) are common in most rivers inhabited by butterfly peacock bass. Many times, while fishing on the Cinaruco, Negro, and Branco rivers, we have experienced attacks by freshwater dolphins on peacock bass on the end of a line or following release. Dolphins are often seen in close proximity to fishing boats, waiting to attack a thrashing peacock bass. One of us (CGM) observed a giant river otter feeding on an Orinoco butterfly peacock bass in the Ventuari River. Juvenile peacock bass are vulnerable to predation by a variety of piscivorous fish, including larger peacock bass, and birds found in the Neotropics. Peacock bass of all sizes are commonly observed with fins damaged by piranhas, and such attacks may occur while fish struggle on the end of a fishing line.

Spectacled caiman (*Caiman crocodilus*) are common in most habitats where Orinoco butterfly peacock bass occur. *Photo: K. Winemiller.*

Growth

Based on size-at-age data across all age classes, Orinoco butterfly peacock bass in the Cinaruco River were estimated to grow an average of 2.85 cm per year. Fish were estimated to grow rapidly, 2.2 cm per month, during their first year of life [8]. Body condition and visceral fat deposits of Orinoco butterfly peacock bass in both the Cinaruco and Ventuari rivers were greatest during the late wet season (December) when the annual flood pulse was beginning to subside, and values were lowest during the late dry season (April–June) when fish were reproducing just prior to the onset of the wet season and rising water levels.

Based on analysis of annual growth rings in otoliths (mineral deposits in the skull that play a role in sensing balance and acceleration), the maximum age of *C. orinocensis* in a sample from the Cinaruco River was estimated to be 3 years, and fish from the Pasimoni River (blackwater tributary of the Casiquiare) was estimated to be 7 years [8]. This difference was inferred to be associated with greater fishing pressure on the Cinaruco stock rather than differences in longevity.

In rivers, *C. orinocensis* grows slower and attains smaller maximum sizes than *C. temensis*, but in some reservoirs where the two have been stocked, *C. orinocensis* sometimes grows larger.

Maximum lengths and weights reported for Orinoco butterfly peacock bass in Venezuela [8] are as follows:

Aguaro River—32 cm, 0.8 kg
Cinaruco River—43.5 cm, 2 kg
Upper Orinoco—45.5 cm, 2.6 kg
Guri Reservoir—56.5 cm, 4.1 kg
Las Majaguas Reservoir—41 cm, 1.9 kg
Modulos de Apure (ponds)—37 cm, 1.3 kg

It should be noted that the large sizes attained by some Guri fish likely are explained by hybridization that has occurred with *C. temensis* [2]. During the time when the above-mentioned

Orinoco butterfly peacock bass from the Atabapo River, blackwater tributary of the Upper Orinoco in southern Venezuela. *Photo: S. Willis.*

surveys were conducted in the Aguaro River and Las Majaguas Reservoir, fish size already had declined due to fishing pressure.

The world record size for *Cichla orinocensis* reported by the IGFA is 67 cm and 7.31 kg (16 lbs, 10 oz) caught in the Urubaxi River, a tributary of the Rio Negro in Brazil.

Population abundance and structure

The density and size structure of peacock bass populations vary according to environmental conditions and fishing pressure. Best approximations of population density and structure of Orinoco butterfly peacock bass in rivers come from studies conducted in the Venezuelan Llanos. Using the mark and recapture method, the abundance of *C. orinocensis* was estimated at 148 fish per hectare in Laguna Larga, a lake connected to the Cinaruco River, and 178 fish per hectare in Laguna Brava, a lake in the floodplain of the Rio Capanaparo [16]. Angling estimates for these lakes showed a higher average catch rate in Laguna Brava (2.1 fish per person per hour) than Laguna Larga (1.65 fish per person per hour). In the Aguaro River, the density of *C. orinocensis* was estimated at 312 fish per hectare [17]. In other locations along this river, including small tributary streams, the density of *Cichla* was relatively low.

The average size of Orinoco butterfly peacock bass from the several locations in Venezuela was reported as follows [8]:

Aguaro River—24 cm, 0.3 kg
Cinaruco River—30.5 cm, 0.7 kg
Upper Orinoco—36 cm, 1.1 kg
Guri Reservoir—36 cm, 1.2 kg
Las Majaguas Reservoir—32 cm, 0.9 kg
Modulos de Apure (ponds)—27 cm, 0.6 kg

Based on these limited samples, it appears that fish in the upper Orinoco, which includes the Ventuari, Casiquiare, and Pasimoni rivers, and Guri Reservoir tend to be larger than fish in rivers, reservoirs, and ponds in the Llanos. This trend likely is influenced by the degree of fishing pressure, because, for example, the Aguaro River is subjected to very heavy fishing pressure while the Upper Orinoco receives very light pressure.

As described at the beginning of this chapter, the Orinoco butterfly peacock bass seems well adapted to environmental conditions in Las Majaguas Reservoir. During the mid-1980s, *C. orinocensis* in Las Majaguas were large, including a 55-cm and 5-kg specimen caught in 1984, as recounted at the beginning of this chapter. Over the next decade, the population of *C. orinocensis* declined dramatically, probably due to the combination of overfishing and reduction of ecosystem productivity that generally accompanies reservoir maturation.

Reproduction

Similar to most peacock bass, *C. orinocensis* generally undergoes a surge in spawning activity near the end of the dry season just before the onset of the rainy season, although some populations may have some individuals spawn during any given period [5, 8]. Spawning tends to be more synchronous in the Llanos region of the Orinoco Basin where there is a very distinct and regular annual wet (June–November) season, and spawning tends to be less synchronous in the Upper Orinoco and Rio Negro where rainfall is less predictable and spread out more evenly over the year.

In the Cinaruco River, the average size at first maturation for *C. orinocensis* was estimated at 31 cm for males and 27 cm for females [8]. For Ventuari fish, the average size of first maturation was 28 cm for both sexes, with the minimum sizes of mature fish reported as 22 cm for males and 21 cm for females [12].

The body condition and fat stores of *C. orinocensis* from the Cinaruco River were lowest during April–June, which coincided with maturation of the gonads in preparation for spawning and then followed by nesting and brood guarding [6]. The dark rings observed in otoliths were formed during this period of reduced feeding and growth [8]. By the end of the rains in September, gonads of both males and females had regressed. Gonad maturation occurred gradually during the period of flood water recession (December–May), and nesting was observed by the end of the dry season. Orinoco butterfly peacock bass in the Aguaro River reveal a similar pattern of gonad development and spawning. Many fish enter floodplain habitats for spawning during May–June, and by July fry can be observed in flooded savannas, creeks, and lakes [18].

In the Ventuari River (Upper Orinoco), some butterfly peacock bass may spawn as early as January, but most fish show maturing gonads from January to March and spawn during the

A healthy Orinoco butterfly peacock bass from the Cinaruco River in the Venezuelan Llanos. *Photo: C. Montaña.*

falling-water period during April–May [12]. By June, most fish have spent gonads, indicating that they have already spawned.

The average number of mature eggs in mature ovaries (batch fecundity) of *C. orinocensis* from the Ventuari River was 3647 [12], with fecundity ranging from 245 to 4643 and larger females producing more eggs. Batch fecundity of *C. orinocensis* from Guri Reservoir was reported as 9900 eggs [19].

Genital papilla, or ovipositor, from which eggs are deposited. *Photo: C. Montaña.*

We have observed *C. orinocensis* pairs on their nests and nests with eggs in rivers and lakes throughout the species' geographic distribution. Nests were always in shallow water and often near logs or rocks that presumably provide cover for the parents. Eggs may be deposited on the surface of logs, rocks, or depressions constructed in sand. The area is always cleaned by the pair by sweeping with the caudal fin and biting or sucking larger pieces of debris. Nest-tending fish are easily frightened and flee from their nest when approached by humans from above or below the water surface; however, when fry become free swimming, the parents are more aggressive and are hesitant to abandon their brood. Brood-guarding parents oftentimes hold their position when approached by a swimmer, often feigning attacks or even dashing in to give a nip. Anglers frequently catch peacock bass that are brood guarding and thus prone to striking lures out of aggression rather than hunger.

Eggs of *C. orinocensis* deposited on a submerged log. When the eggs hatch, the larvae are transported by the parents to pits constructed in loose substrates nearby. *Photo: K. Winemiller.*

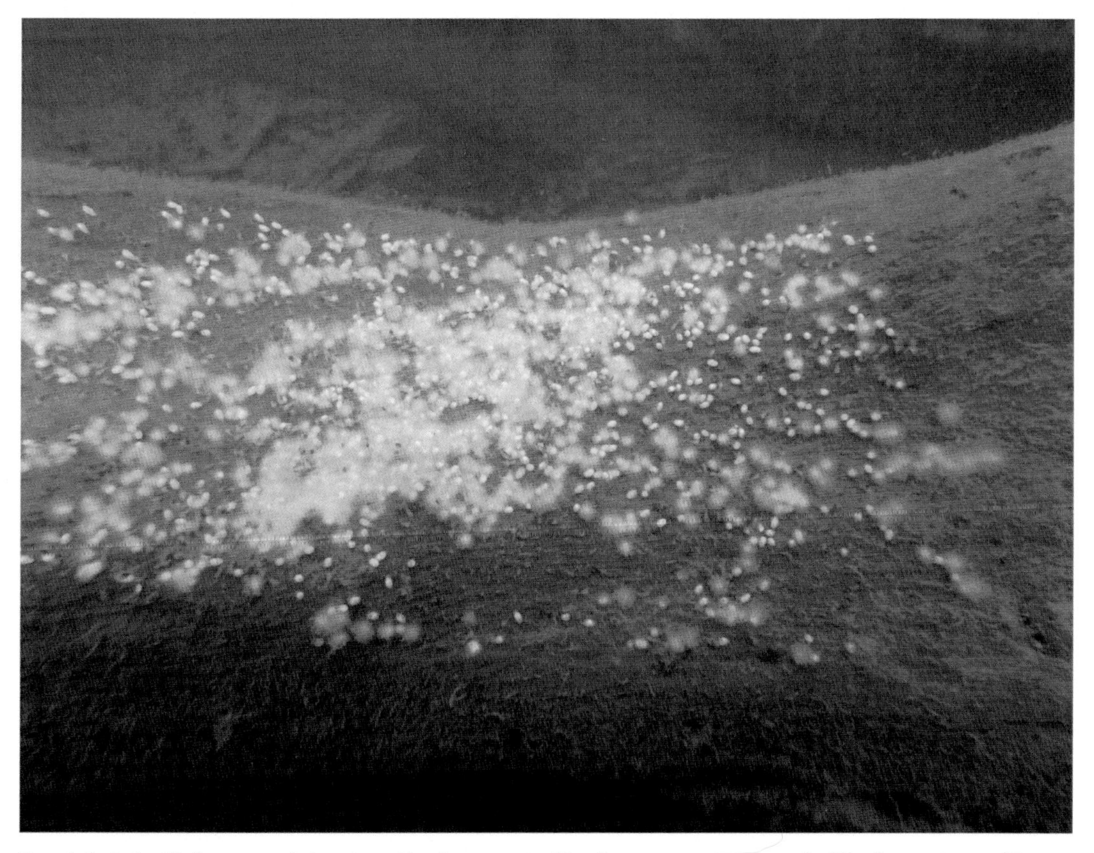

Eggs infected with fungus and abandoned by the parents either because most were unfertilized or water quality was insufficient to support development. *Photo: K. Winemiller.*

A school of *C. orinocensis* fry in the clear water of Agua Boa, a creek in the floodplain of the Rio Negro. *Photo: K. Alofs.*

References

[1] S.O. Kullander, E.J.G. Ferreira, A review of the South American cichlid genus *Cichla*, with descriptions of nine new species (Teleostei: Cichlidae), Ichthyol. Explor. Freshw. 17 (2006) 289–398.

[2] S.C. Willis, M. Nunes, C.G. Montaña, I.P. Farias, G. Ortí, N. Lovejoy, The Casiquiare River acts as a dispersal corridor between the Amazonas and Orinoco River basins: biogeographic analysis of the genus Cichla, Mol. Ecol. 19 (2010) 1014–1030.

[3] S.C. Willis, J. Macrander, I.P. Farias, G. Orti, Simultaneous delimitation of species and quantification of interspecific hybridization in Amazonian peacock cichlids (genus *Cichla*) using multi-locus data, BMC Evol. Biol. 12 (2012) 96, https://doi.org/10.1186/1471-2148-12-96.

[4] S.C. Willis, I.P. Farias, G. Orti, Multi-locus species tree for the Amazonian peacock basses (Cichlidae: *Cichla*): emergent phylogenetic signal despite limited nuclear variation, Mol. Phylogenet. Evol. 69 (3) (2013) 479–490.

[5] K.O. Winemiller, Ecology of peacock cichlids (*Cichla* spp.) in Venezuela, J. Aquaric. Aquat. Sci. 9 (2001) 93–112.

[6] D.B. Jepsen, K.O. Winemiller, D.C. Taphorn, Temporal patterns of resource partitioning among *Cichla* species in a Venezuelan blackwater river, J. Fish Biol. 51 (1997) 1085–1108.

[7] K.O. Winemiller, D.C. Taphorn, A. Barbarino-Duque, The ecology of *Cichla* (Cichlidae) in two blackwater rivers of southern Venezuela, Copeia 1997 (1997) 690–696.

[8] D.B. Jepsen, K.O. Winemiller, D.C. Taphorn, D. Rodriguez Olarte, Age structure and growth of peacock cichlids from rivers and reservoirs of Venezuela, J. Fish Biol. 55 (1999) 433–450.

[9] C.G. Montaña, C.A. Layman, K.O. Winemiller, Gape size influences seasonal patterns of piscivore diets in three Neotropical rivers, Neotrop. Ichthyol. 9 (2011) 647–655.
[10] D. Rodríguez-Olarte, D.C. Taphorn, Ecologia trofica de *Cichla orinocensis* Humbolt 1833 (Pisces, Teleostei, CIchlidae) en un humedal de los llanos centrales de Venezuela, Biollania 13 (1997) 139–163.
[11] J.D. Williams, K.O. Winemiller, D.C. Taphorn, L. Balbas, Ecology and status of piscivores in Guri, an oligotrophic tropical reservoir, N. Am. J. Fish. Manag. 18 (1998) 274–285.
[12] C.G. Montaña, D.C. Taphorn, C.A. Layman, C. Lasso, Distribución, alimentación and reproducción de tres species de pavones *Cichla* spp. (Perciformes, Cichlidae) en la cuenca del Rio Ventuari, Venezuela, Mem. Fund. La Salle Cienc. Nat. 165 (2007) 83–102.
[13] A. Barbarino, Diagnóstico del recurso pesquero como base para su reglamentación en el embalse Las Majaguas, estado Portuguesa, Venezuela (Tesis de Maestría), Universidad de los Llanos Occidentales Ezequiel Zamora, Guanare, Venezuela, 1996.
[14] O. Castillo, J. Carillo, R. Rojas, D. Urdaneta, J. Peñuela, R. Barrios, T. Jurado, L. Velásquez, M. Piñero, Contribución al conocimiento de la ictiofauna del Embalse Masparro, estado Barinas, Unversidade del Los Llanos Occidentales Ezequiel Zamora, Guanare, Venezuela, 2009.
[15] D.F. Novoa, J. Koonce, F. Ramos, La ictiofauna del lago Guri: composición, abundancia y potencial pesquero. II. Evaluación del potencial pesquero del lago Guri y estrategias de ordenamiento pesquero, Mem. Soc. Cienc. Nat. La Salle 49 (1989) 159–197.
[16] D.C. Taphorn, A. Barbarino, Evaluacion de la situacion actual de de los pavones (*Cichla* spp.) en el Parque Nacional Capanaparo-Cinaruco, Estado Apure, Venezuela, Nature 96 (1993) 10–25.
[17] D. Rodríguez-Olarte, D.C. Taphorn, Aspectos de la ecología reproductiva del Pavón Estrella *Cichla orinocensis* Humboldt 1833 (Pisces: Ostariophysi: Cichlidae) en el Parque Nacional Aguaro-Guariquito, Venezuela, Mem. Fund. La Salle Cienc. Nat. 161–162 (2005) 5–17.
[18] D. Rodriguez-Olarte, D.C. Taphorn, Distribucion y abundancia de pavones (*Cichla orinocensis* y *C. temensis*, Pisces:Cichlidae) en un humedal de los Llanos Centrales en Venezuela, Vida Silv. Neotrop. 8 (1–2) (2001) 43–50.
[19] C. Lasso, A. Machado-Allison, R.P. Hernández, Consideraciones zoogeográficas de los peces de la Gran Sabana (Alto Caroní) Venezuela, y sus relaciones con las cuencas vecinas, Mem. Soc. Cienc. Nat. La Salle 49–50 (1989) 109–129.

CHAPTER

4

Royal peacock bass, *Cichla intermedia* (Machado-Allison, 1971)

Cichla intermedia

Origin of name

Latin "*intermedia*" meaning "occurring in a middle position"; referring to the black lateral band

Synonyms

Cichla nigrolineatus, Ogilvie 1966

Common names

Royal peacock bass, Pavón real, Blackstripe peacock bass

Geographic distribution

Cichla intermedia is endemic to the Orinoco and Casiquiare river basins.

Natural history

Restricted to rivers with clear water (not tea-stained) with few suspended particulates or moderate black water with pH from 5.5–6. Encountered at greater depths in areas of river channels where the water is flowing and there is abundant submerged structure in the form of rocks, logs or brush. Primarily piscivorous.

https://doi.org/10.1016/B978-0-323-85157-2.00011-1

It was November of 2004, and I (CGM) was making my first exploration of the Caura River in Venezuela's Bolivar State. Accompanied by my colleague Stuart Willis, the leading expert on peacock bass genetics, I arrived at the Port of Maripa to meet Felix Daza, a biologist working for the Wildlife Conservation Society. Felix was in the area conducting research on the morocoto (*Piaractus brachypomus*), a large fruit-eating fish belonging to the same family as the piranhas (Serrasalmidae). The morocoto is the mainstay of that region's subsistence and commercial fisheries. Felix had been monitoring morocoto using radiotelemetry and also compiling a list of the region's fish species, so he was very familiar with the river and its fish and local people. Moreover, Felix was assisted by two fishermen from one of the Yekwana communities who knew the river and fishes very well.

We started our river journey on a hot, sunny afternoon. We packed our provisions and fishing gear and boarded a 20-m wooden boat driven by José, a local fisherman from Maripa. Our goal was to collect peacock bass for Stuart's genetics research and for my ecological studies. Three species of *Cichla* are reported for the Caura River, *C. orinocensis*, *C. temensis* and *C. intermedia*, the royal peacock. We were particularly interested in *C. intermedia*, because this species inhabits clearwater rivers in the Llanos and upper Orinoco regions, but it also has a healthy population in the Caura River which drains the Guiana Shield in southern Venezuela and has tea-stained, acidic waters. We wondered if this population might be genetically divergent, or perhaps even be a separate species.

It was the dry season and the river's receding waters had exposed rocky shoals around which José deftly maneuvered our boat. Those rocky shoals also are prime habitat for royal

Royal peacock bass (*C. intermedia*) caught with a jig amid rocks in the Venezuela's Caura River. This large male had a huge nuchal hump indicative of the spawning season. *Photo: C. Montaña.*

peacock bass. To capture fish specimens, we brought fishing rods, gillnets and harpoons. The fishermen told us that most peacock bass are caught with harpoons at night when the fish come near shore to sleep.

On the first day, we caught some fish in our gillnets, but none were peacock bass. The second day, we moved upstream from the village of Aripoa, stopping along the way to fish in rocky areas using lures of various types as well as some live bait. Later that afternoon near Trincheras, a small Yekwana village, we got out of the boat and hopped from rock to rock with fishing rods in hand. I tied on a silver and white jig, a heavy lure that would work well in the rapid water swirling around the rocks. In a matter of minutes, I hooked a royal peacock bass—a huge male with a massive nuchal hump.

After nightfall, we fished in the rocks with harpoons and collected over 20 royal peacock bass of various sizes. We photographed and measured the specimens, collected tissue samples, and preserved several for later study. Most fish had brilliant colors, and the males had huge nuchal humps, indications that these fish were in the midst of their spawning season. Over the next several days, we observed many fish on nests scattered among the rocks. The royal peacock bass in the Caura are darker green, deeper-bodied, and most were larger than those we collected from the Cinaruco, Ventuari, and Casiquiare rivers.

Male *C. intermedia* with large nuchal hump that was captured at night using a harpoon. *Photo: C. Montaña.*

The royal peacock bass only occurs in river channel habitats with flowing water, and usually where there are lots of rocks. This probably explains why relatively few anglers visiting locations in the Orinoco Basin manage to catch a royal peacock bass. Most international anglers who visit the region target the large speckled peacock bass (*C. temensis*), and therefore they concentrate their efforts on floodplain lakes and channel habitats with sluggish current, often fishing with surface lures. This is unfortunate, because the royal peacock bass is one of the most beautiful species and also one of the strongest fighters.

Historically, the people who live along the Caura guarded the river from illegal fishing by outsiders, and they only harvest what they needed for subsistence. In recent years, there has been greater incentive to supply fish to larger markets, leading to greater fishing pressure. Deforestation and illegal mining for gold and diamonds also have impacted the region's ecology. Due to the difficult economic and security situation in Venezuela over the past decade, there hasn't been much research done on the Caura or other rivers in the country. Time will tell if the next scientific survey of the Caura will find a healthy population of this beautiful fish.

Fisherman with a royal peacock bass from the Caura River. This nesting male shows extreme development of the nuchal hump. *Photo: C. Montaña.*

Identification

The royal peacock bass is distinguished from all other species by the presence of a black lateral band composed of a series of interconnected irregular blotches that extend the length of the body from the edge of the operculum into the caudal peduncle. This band is discontinuous in adults and continuous in juveniles. In adults, the horizontal blotches are often outlined by white, cream, yellow or light green coloration. Usually 5–6 of the more anterior blotches are larger than the other blotches and spots that form the lateral band. Each of these larger blotches is intersected by a dark vertical bar that extends from the top of the body toward the belly, and in some individuals additional vertical bars are present. The intensity of vertical bar expression depends on the fish's environment and physiological state, but from 5 to 10 bars are always evident. In juveniles (<10 cm), the black lateral band is more uniform and extends into the postorbital region of the head, and the vertical bars tend to be fainter and may not extend as far above and below the blotches.

Depending on environmental conditions and reproductive state, the background coloration of the head and body varies from dull gray to yellow to green to a brilliant turquoise-green.

Sometimes the dorsal region is gray or green while the belly region is yellow or yellow-green. The belly region is usually white or a lighter shade of the body background coloration. In some individuals, the ventral half of the head and body is yellow. Adults usually have several small black spots or blotches in the postorbital region, cheeks, and/or gill cover. In some adults, the branchiostegal membranes are bright orange. The ring encircling the caudal ocellus is usually white, cream or very light yellow rather than bright yellow or gold as in some other peacock bass species.

The dorsal fin and upper half of the caudal fin are gray and greenish-gray with light blue or blue-green spots and streaks. These spots are translucent in the fins of juveniles. The pelvic, anal and lower half of the caudal fin usually are orange, but sometimes this is masked by green pigmentation, especially in breeding adults of some populations. The caudal fin and rayed portion of the dorsal fin sometimes have a brilliant turquoise border. The eye

Cichla intermedia from the Cinaruco River showing nine vertical bars and spots along the midline. *Photo: K. Winemiller.*

Royal peacock bass from the Ocamo River, clearwater tributary of the Upper Orinoco in Venezuela. *Photo: S. Willis.*

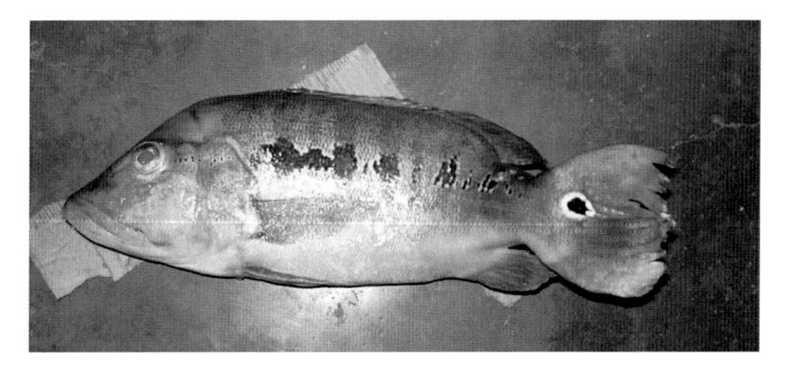

A robust specimen of *C. intermedia* from the Atabapo River, a blackwater tributary of the Upper Orinoco on the border between Colombia and Venezuela. This species is more commonly encountered in clearwater rivers, and this individual may have strayed into the Atabapo from the Orinoco. *Photo: C. Montaña.*

Immature royal peacock bass from the Cinaruco River. *Photo: C. Montaña.*

iris is always red, even in subadult and non-reproductive adults. In breeding fish, the iris turns a brilliant crimson that is more intense and beautiful than any other peacock bass species.

Juveniles smaller than 10 cm are similar to those of *C. temensis* and *C. melaniae*, with a dark lateral stripe extending from the eye to a spot at the base of the caudal fin. Juveniles also show 6–8 vertical bars.

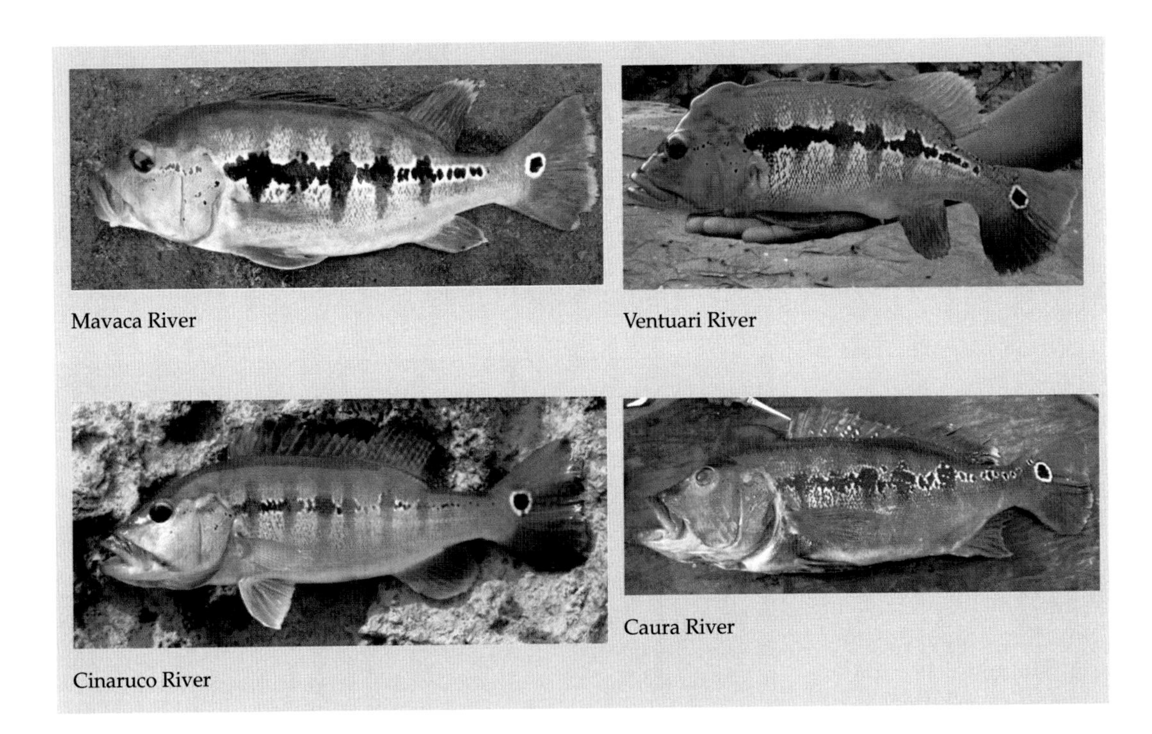

Mavaca River

Ventuari River

Cinaruco River

Caura River

Distribution and habitat

The royal peacock bass is endemic to the Orinoco River Basin in Colombia and Venezuela. The species is not found in the Amazon Basin, despite the existence of a corridor for dispersal via the Casiquiare River, which flows from the upper Orinoco to the upper Rio Negro. The Casiquiare originates as a channel that branches from the upper Orinoco in southern Venezuela (Amazonas state) and flows southwest to join the upper Rio Negro, the immense blackwater tributary of the Amazon River. The royal peacock bass apparently is restricted to rivers with clear water (not tea-stained) with few suspended particulates and moderate black water with pH from 5.5–6 [1]. The species appears to be intolerant of extreme blackwater conditions, such as those found in the lower Casiquiare River and Rio Negro (pH 4–5), and consequently it appears unable to invade the Amazon Basin. However, there is a native population of *C. intermedia* that has adapted to conditions in the Caura River, a blackwater tributary of the middle Orinoco that drains the Guiana Shield formation in

Distribution of the royal peacock bass (*Cichla intermedia*) in Colombia and Venezuela.

Venezuela's Bolivar state. This population of royal peacock bass displays some differences in coloration (uniform dark green background coloration of head, body and fins) and also seems to grow larger than fish from other regions, although this observation could be affected by differences in local fishing pressure. A population of *C. intermedia* is reported to inhabit a tributary of the Caquetá River near the town of San José del Fragüa in southern Colombia. The Caquetá is a tributary of the Amazon, and we have been told that this non-native population became established when fish escaped from an ornamental fish farm during a flood.

The royal peacock bass generally is encountered in areas within river channels where the water is flowing and there is abundant submerged structure in the form of rocks, logs or brush [1–4]. Its strong affinity for structured habitats and water current accounts for its ability to partition habitat with *C. temensis* and *C. orinocensis*, two species that often coexist with *C. intermedia* but inhabit both floodplain lakes and river channels, generally in areas with slow flowing or still water. We have observed nesting pairs of royal peacock bass in backwater areas lacking flow, but these habitats apparently are only used by adults for nesting. Juvenile *C. intermedia* have been observed and captured along the shoreline of

Typical habitat of *C. intermedia* in the Ocamo River, Upper Orinoco Basin, Venezuela. *Photo: K. Winemiller.*

rivers where water flow was slow, but we have never collected juveniles from floodplain habitats. Royal peacock bass tend to stay near the bottom and are less likely than other peacock bass species to ascend in the water column to attack lures. We have found that the most reliable lures for capturing the royal peacock bass are jigs, spoons, and heavy inline spinners that can be fished deep, particularly when retrieved in the same direction as the water flow.

The catch from less than an hour fishing on the Ocamo River in the Upper Orinoco. In addition to royal peacock bass, anglers caught black piranhas (*Serrasalmus rhombeus*), a large matrinchã (*Brycon amazonicus*), and a dogtooth characin (*Hydrolycus tatauaia*). *Photo: K. Winemiller.*

Feeding

Most of the available information on the diet of the royal peacock bass is derived from research we conducted with colleagues on the Cinaruco River, a clearwater tributary of the middle Orinoco located in the Llanos region of Venezuela [2]. One of us (CGM) also conducted research with colleagues on the ecology of *C. intermedia* in the Ventuari River, a clearwater tributary of the upper Orinoco in the rainforest of Venezuela's Amazonas state [4]. The diet of *C. intermedia* was dominated by fish in both rivers, with 82% of fish sampled from the Cinaruco having consumed fish, and 100% of the fish sampled from the Ventuari having fish remains in their stomachs. Only fish larger than 20 cm were examined in these studies, and only three specimens from the Ventuari contained food items, and all were characiform fish (*Brycon*, *Hemigrammus* and *Moenkhausia* spp.). The principal types of fish ingested by royal peacock bass in the Cinaruco River were catfish, cichlids, tetras and other characiform fishes [1, 2]. Many of the fish remains that were recovered from stomachs could not be identified to species or family, and the sample sizes were fairly small. Royal peacock bass in the Cinaruco also consumed shrimp (13% by volume) and aquatic insects (7.7% by volume).

The Cinaruco River experiences a large annual flood pulse, and *C. intermedia* revealed very low diet overlap with both *C. temensis* and *C. orinocensis* during the periods of rising and falling water levels. During the period of lowest water levels, when the floodplains were dry and fish were confined to channel and lake habitats, *C. intermedia* had high diet overlap with *C. orinocensis* (90%) and *C. temensis* (69%). *Cichla intermedia* and *C. orinocensis* were similar

Royal peacock bass caught from a shallow area along the bank of the Cinaruco River. This pair presumably was defending a nest and attacked the lures out of aggression rather than for feeding. *Photo: K. Winemiller.*

sizes and the average size of prey recovered from stomachs was similar for these two species, except for the period of falling water levels when *C. orinocensis* consumed larger prey [2]. In contrast, *C. temensis* was larger on average and also consumed larger prey than the other two peacock bass species. The exception was during the rising water period, when all three species consumed a similar range of prey sizes. Unlike *C. temensis* and *C. orinocensis* that revealed a decline in average size of prey consumed throughout the period of floodwater recession, *C. intermedia* revealed virtually no change in the average and range of prey sizes consumed during various times of the year.

Growth

By analyzing growth rings on otoliths (again, these are mineral deposits in the skull involved in sensing balance and acceleration), it was estimated that *C. intermedia* in the Cinaruco River grew about 2 cm per month [5]. Assuming that annual growth rings are associated with

A healthy royal peacock bass from the Ocamo River, a tributary of the Upper Orinoco in Venezuela. *Photo: C. Montaña.*

intervals of growth between spawning bouts that occur during the end of the annual dry season (i.e., corresponding to the low-water period of the annual flood pulse), the maximum documented age of *C. intermedia* in the Cinaruco was 6 years. The minimum size at maturation was 32 cm for males and 27 cm for females. The largest *C. intermedia* from a survey of fish from the Cinaruco River was 42.9 cm weighing 1.6 kg, and the largest specimen from a survey of the Casiquiare in southern Venezuela (Amazonas state) was 40 cm and 1.5 kg.

The world record royal peacock bass reported by IGFA weighed 3.86 kg (8 lb, 8 oz) and was caught in the Villacoa River, a clearwater tributary of the Orinoco in Venezuela's Bolivar state.

Population abundance and structure

Studies estimating population abundance of royal peacock bass are lacking, but ecological studies conducted in the Cinaruco, Casiquiare and Ventuari rivers in Venezuela consistently yield fewer specimens of *C. intermedia* than *C. temensis* or *C. orinocensis*. The lower apparent abundance of the royal peacock bass in these rivers probably is due to its restrictive habitat requirements and the difficulty of sampling fish that tend to occupy zones near the substrate in areas with flowing water, rocks and logs. However, in the Caura River, a blackwater tributary of the Orinoco in Venezuela's Bolivar state, *C. intermedia* appears to be more abundant than the other two species. This is probably because the Caura's relatively fast-flowing water and rocky substrate create conditions more favorable for the royal peacock bass. Moreover, the river's narrow floodplains and scarcity of lakes over much of its course probably explain why there are fairly small stocks of *C. temensis* and *C. orinocensis*.

The average size of royal peacock bass in samples from the Casiquiare and Ventuari rivers (both averaged 33.2 cm, 0.8 kg) was slightly larger than the average size of fish from the Cinaruco River (31 cm, weight 0.7 kg) [2, 4]. The size range of *C. intermedia* specimens was greater for the Cinaruco (20.5–42.9 cm and 0.2–1.6 kg) as compared to the Casiquiare (29.5–40 cm and 0.5–1.5 kg).

A mark-recapture study conducted on the Cinaruco River from 1999 to 2003 tagged 142 royal peacock bass and obtained only 6 recaptures [6]. Fish smaller than 1 kg had moved an average of 0.2 km, and fish weighing greater than 1 kg had traveled an average of 0.3 km. These distances were slightly less than those recorded for *C. orinocensis* (0.3–0.7 km),

Cichla intermedia from Venezuela's Cinaruco River. *Photo: K. Winemiller.*

and much less than those recorded for *C. temensis* (0.8–6.9 km) in the same study. Thus, it seems that *C. intermedia* has greater site fidelity than the other two *Cichla* species. Only a small percentage of tagged fish were recaptured, which may be attributed to the large size of the populations, mortality, and/or tag loss.

Reproduction

We have observed several pairs of *C. intermedia* on their nests in the Cinaruco, Caura and Ocamo rivers. Nests were located in very shallow water (less than 1 m) on rocks within the main river channel. Large boulders sheltered nest locations from strong current. Pairs were reluctant to abandon their nests, usually remaining until approached on foot or by boat at a distance of 2–3 m.

Royal peacock bass on its nest in a rocky shoal of the Ocamo River in southern Venezuela. *Photo: K. Winemiller.*

Like other peacock bass, *C. intermedia* apparently reduces or ceases feeding while on their nest and brood guarding [2]. Cinaruco fish revealed the highest proportion of empty stomachs near the end of the dry season (April–June) just before the water began to rise again. This was the same period that gonads of most adults were maturing [5].

Specimen of *C. intermedia* captured from rocky shoals of the Caura River, Venezuela. *Photo: C. Montaña.*

References

[1] K.O. Winemiller, D.C. Taphorn, A. Barbarino Duque, The ecology of *Cichla* (Cichlidae) in two blackwater rivers of southern Venezuela, Copeia 1997 (1997) 690–696.

[2] D.B. Jepsen, K.O. Winemiller, D.C. Taphorn, Temporal patterns of resource partitioning among *Cichla* species in a Venezuelan blackwater river, J. Fish Biol. 51 (1997) 1085–1108.

[3] K.O. Winemiller, Ecology of peacock cichlids (*Cichla* spp.) in Venezuela, J. Aquaric. Aquat. Sci. 9 (2001) 93–112.

[4] C.G. Montaña, D.C. Taphorn, C.A. Layman, C. Lasso, Distribución, alimentación and reproducción de tres species de pavones *Cichla* spp. (Perciformes, Cichlidae) en la cuenca del Rio Ventuari, Venezuela, Mem. Fund. La Salle Cien. Nat. 165 (2007) 83–102.

[5] D.B. Jepsen, K.O. Winemiller, D.C. Taphorn, D. Rodriguez Olarte, Age structure and growth of peacock cichlids from rivers and reservoirs of Venezuela, J. Fish Biol. 55 (1999) 433–450.

[6] D.J. Hoeinghaus, C.A. Layman, D.A. Arrington, K.O. Winemiller, Movement of *Cichla* species (Cichlidae) in a Venezuelan floodplain river, Neotropical Ichthyology 1 (2003) 121–126.

CHAPTER

5

Speckled peacock bass, *Cichla temensis* (Humboldt, in Humboldt & Valenciennes, 1821)

https://doi.org/10.1016/B978-0-323-85157-2.00013-5

Cichla temensis

Origin of name

Latin "*temensis*" meaning "originating from Temi River"

Synonyms

Cichla atabapensis (Humboldt, in Humboldt & Valenciennes 1821), *Cichla tucunare* (Heckel 1840), *Cychla flavo-maculata* (Jardine 1843), *Cychla trifasciata* (Jardine 1843), *Cichla unitaeniatus* (Magalhães 1931)

Common names

Speckled peacock bass, Banded peacock bass, Three-barred peacock bass, Pavón cinchado, Pavón lapa, Pavón pinta lapa, Pavón trucha, Tucunaré açu, Tucunaré lapa.

Geographic distribution

Cichla temensis is restricted to the Rio Negro, Orinoco and lower Madeira basins.

Natural history

Common in river channels and floodplain lakes near shore in areas with submerged sticks and logs lacking strong current. Sometimes located in deeper water further from the shoreline where it waits for prey to venture away from cover near shore. Primarily piscivorous.

It was December 1992, and our family was spending the Christmas holiday at the house of Leslie's parents in Arlington, Texas, when I (KW) came across an article in the *Dallas Morning News* about two local men who had been invited to explore a remote region of Venezuela to fish for giant peacock bass. At that time, I had already spent years studying fish ecology in Venezuela and several other tropical countries. Most of my Venezuelan work had been done in the Llanos region, and the newspaper article's account of a river teeming with huge peacock bass seemed almost unbelievable. The featured river, the Pasimoni, is located in a rainforest region of Venezuela's Amazonas state. It was a pristine environment that supported an untouched native population of peacock bass. This scenario is an exception in today's world where there are few unspoiled places. As a fisheries biologist, I was aware that an unexploited population was indeed likely to be dominated by large fish. I had to see this place!

As luck would have it, I already had been approved for a research grant to do a project in the Venezuelan Llanos during the first 2 months of the new year. After completing that work, I consulted my friend and colleague, Aniello Barbarino, about the feasibility of using the remaining research funds to explore the Pasimoni and its population of giant peacock bass. Aniello and I consulted some colleagues at the University of the Western Llanos about the steps required to secure government permission for conducting scientific research in this remote area. The Pasimoni River is a tributary of the Casiquiare River near Venezuela's border with Colombia and Brazil and is inhabited by indigenous groups who have government protection from outside interference. We also were well aware that logistics would be challenging. Not only would we need to arrange a flight in a small aircraft to the military airstrip at the tiny village of San Carlos del Rio Negro, but we would also have to obtain official permits to enter the region, conduct research, and obtain gasoline. Furthermore, we had to transport all of our provisions including a boat motor, camping supplies, fishing gear, research instruments, and containers with chemicals for preserving fish specimens. Aniello was skeptical that all of this was feasible, but thanks to his negotiating skills and a little bit of luck, we managed to accomplish what at first appeared to be an impossible feat.

The air-taxi flight from Puerto Ayacucho to San Carlos del Rio Negro provided breathtaking views of endless rainforests, blackwater rivers, rapids, and tepuis (ancient, weathered mesas of the Guiana Shield), and also literally took our breath away because of the inebriated state of our pilot. *En route*, the plane made a stop at San Fernando de Atabapo where some local people managed to force some strong coffee into our pilot after observing his bloodshot eyes and slurred speech. Upon arrival in San Carlos del Rio Negro, we presented our permits to the local military commander, and then located a young man from the village who could take us in his dugout canoe up the Rio Negro, into the Casiquiare River, and then up the Pasimoni. With only a few days available for us to work in the area, our goal was to catch as many peacock bass as possible in order to conduct age and growth analyses. These data would provide a rare opportunity to compare the age and size structure of an unfished stock with those in other regions that have been exploited for generations.

Our guide must have thought we were either crazy, stupid, or both, because we had ample fishing and research gear, but minimal food with only the clothes on our backs plus sleeping hammocks. Fortunately, I had a plastic tarp to cover my hammock, whereas Aniello had to improvise by hanging some plastic trash bags over his hammock for protection from the nightly torrential rains. It took us an entire day to motor upstream to a suitable site on the Pasimoni River where we could camp by the river with good peacock bass habitat nearby. Our campsite was a granite outcrop at the water's edge, just upstream from a large lake separated from the river channel by a long, narrow spit of sand. Viewing the black water of the Pasimoni against the white sandbank was reminiscent of espresso coffee lapping against white granulated sugar. Given our limited time, we set out early to try to catch some large peacock bass since I was especially interested in estimating the ages of giant specimens like the ones that I'd read about in the newspaper. We spent an entire morning and most of the afternoon casting from the boat, but with few results—only a few pike characins (*Acestrorhynchus falcirostris*), piranhas (*Serrasalmus* species), and a couple of small Orinoco butterfly peacock bass (*C. orinocensis*). We encountered a boat with two anglers from Alabama, Tom and Amy Nash, who were camped downstream with their guides. They explained that the big speckled peacocks were not biting because it had been raining and the water level had risen. The best conditions for fishing are when the water level is falling. Falling water forces baitfish to move from flooded areas along the shore into deeper water where they are more vulnerable to attack.

The Pasimoni River is an extreme blackwater tributary of the lower Casiquiare. *Photo: K. Winemiller.*

Aniello grew weary of all the casting without results, so he asked to be left on the sandspit to fish using his preferred method—a handline with bait. He used strips of flesh cut from one of the pike characins we'd retained. I continued casting around the lake until dusk and then went to retrieve Aniello and return to camp. When I met Aniello, he asked me how I did, and I replied that I'd caught just one small butterfly peacock. I asked him how he did, and he glumly said "no muy bueno, … los unicos que saqué fueron estos" (not very good, … I only caught these). He leaned down and proudly hoisted up two massive peacock bass. In total he'd caught five *C. temensis*, each weighing 7–9kg. Fishing with lures is good sport, but we needed specimens for research, and Aniello had discovered a foolproof method to catch them. At sunrise we headed back to the sandspit with hooks baited with strips of fish flesh. I used my rod and reel, and Aniello used his handline. We had a friendly competition to see who could catch the most fish, and Aniello soon gained the lead.

I was slowly retrieving my bait in shallow water when I noticed a wake moving behind it. I stopped reeling and twitched the line, and something grabbed it. The fight was on, and I struggled to keep the fish in shallow water near the sandbank where there weren't any submerged sticks. The fish leapt and made a clumsy summersault, landing with a big splash. This was a big fish! I eventually eased it onto the sandbank, held it up, and heard Aniello mutter "grande." He had caught more fish, but clearly, I had caught the largest. Our scale topped out at 10kg, and this fish obviously was much heavier. I didn't have the heart to sacrifice this

Aniello Barbarino with handline catch from the Pasimoni River, 1993. *Photo: K. Winemiller.*

beautiful, huge fish for our research, so we quickly photographed, measured, and released it. Based on its measurements (not only its length, but also its impressive girth), I estimated it weighed over 11 kg (about 25 lbs). Over the next 2 days, we caught more big fish. We measured each specimen and removed their otoliths (those mineral deposits in the skull that form daily growth layers) for age analysis in the lab. Except for one small fish weighing about 2 kg, all of the speckled peacock bass weighed over 7 kg. We had witnessed something quite rare in today's world, an unfished population in a pristine habitat! I suspect that these abundant large fish regulate their own population by cannibalizing smaller fish. We would later determine that the large fish in the Pasimoni were, on average, significantly older than speckled peacock bass in Llanos rivers that have been exploited by people for decades.

In subsequent years, I returned with colleagues to the Pasimoni several times to conduct surveys of regional fish diversity. Our research was aimed at determining if the Casiquiare River, which originates as a branch flowing from the Upper Orinoco eventually connecting with the Upper Rio Negro, serves as a corridor for fish dispersal between the two great river basins, or if the Casiquiare contains ecological barriers preventing dispersal of certain kinds of fish. We ultimately determined that some species require clearwater conditions found in the Upper Orinoco and upper reaches of the Casiquiare, other species are adapted to black-water conditions in the Casiquiare's lower reaches and Rio Negro, and certain species tolerate both kinds of environments [1, 2]. As for the Pasimoni's giant speckled peacocks, sadly, we observed a marked decline in the average and maximum sizes of fish over a period of just a few years. This was a predictable response to an increase in subsistence and sport fishing. In nutrient-poor blackwater rivers, it apparently takes several years to replace the largest, oldest fish when they are removed from the population. Although the stock can be sustained despite a certain amount of fishing mortality, the large "trophy" fish disappear rapidly. The only way to keep big fish in the population is to reduce fishing pressure and practice catch-and-release, although this never eliminates fishing mortality entirely.

Author with his largest speckled peacock bass, Pasimoni River, Venezuela, 1993. This male shows the three-barred pattern typical of fish during low-water periods. *Photo: K. Winemiller.*

Identification

Among all the species of peacock bass, *Cichla temensis* has perhaps the most highly variable coloration pattern. Like most people with limited experience catching or observing this species, the early ichthyologists believed that fish with different coloration patterns represented two or three different species. In fact, these patterns are influenced by age, environmental conditions, and reproductive status. Both males and females undergo a transition in coloration patterns in response to seasonal changes of water level that cue development of gonads and spawning [3]. A recent study that included analysis of morphological and genetic data confirmed that *C. temensis* in the Rio Negro transition from the speckled pattern typical of the high-water season to the three-barred pattern as water levels fall and fish prepare for spawning and brood guarding [4].

Cichla temensis from the Cinaruco River showing the speckled pattern. *Photo: K. Winemiller.*

Cichla temensis from the Pasimoni River showing the three-barred pattern. *Photo: K. Winemiller.*

Cichla temensis from the Ventuari River showing a transitional pattern. *Photo: C. Montaña.*

Cichla temensis from the Cinaruco River showing weak expression of vertical bars. *Photo: K. Winemiller.*

Subadults and nonbreeding adults are bronze or gray-bronze, with variable expression of dark vertical bars 1, 2, and 3, ranging from barely perceptible to black. The bronze background coloration tends to be darker in fish living in extreme black waters. The body may be covered with white or cream spots and dashes arranged in horizontal rows along the sides of the body, including the caudal peduncle. White spotting also is distributed on the head, dorsal fin, and upper half of the caudal fin. Subadult and adult fish have irregular black blotches and spots on the head in the postorbital and opercular (gill cover) area. These blotches and spots are generally bordered in white or cream to present a striking contrast against the bronze background. The throat region, including branchiostegal membranes, and lower portion of the body may be shaded yellow, orange, or orange red. The pelvic, anal, and lower half of the caudal fin are orange or red. The iris of the eye is orange or red in these speckled fish.

A very dark speckled peacock bass from the black water of Guri Reservoir. *Photo: K. Winemiller.*

Subadult *C. temensis* showing the speckled pattern but with some development of the vertical barring pattern. *Photo: H. López-Fernández.*

The three-barred pattern of prebreeding, nesting, and brood-guarding fish consists of head and body background coloration that ranges from yellow to green with distinct vertical bars 1, 2, and 3 that extend from the base of the dorsal fin toward the belly region. The dorsal region is darker than the belly which often is bright orange or orange-red and extends to the throat region. The ventral fins, anal fin, and lower half of the caudal fin also are bright orange or red, although in some environmental conditions they may be yellow or even olive green. The dorsal and upper half of the caudal fin are bluish gray, sometimes with an iridescent blue sheen, and sometimes covered with small translucent white spots. The iris of the eye is brilliant red.

Juveniles (<5cm) have a dark horizontal stripe extending from the eye to the base of the caudal fin. The body is silver-gray and darker in the dorsal region and lighter at the belly. As the fish grows, the background color gradually transitions to gray-bronze or bronze, and vertical bars form at positions 1, 2, and 3. As fish advance to subadult size, the horizontal stripe fades away and a black spot near the base of the caudal fin develops into the caudal ocellus. The dorsal, caudal, and anal fins are transparent and turn gray with translucent spots as the fish grows.

Cichla temensis, one showing the three-barred pattern and one with the speckled pattern, caught by David Hoeinghaus from the same habitat in the Pasiba River, a blackwater tributary of the Casiquiare River in southern Venezuela. *Photo: C. Montaña.*

Juvenile *C. orinocensis* (top) and *C. temensis* (bottom) from the lower Branco River. *Photo: K. Winemiller.*

Distribution and habitat

Cichla temensis is restricted to the Rio Negro, Orinoco, and lower Madeira basins. The species is common and supports important sport fisheries in blackwater habitats of the Rio Negro and its tributaries in Brazil, as well as clearwater rivers of the Orinoco Basin, such as the Bita in Colombia and Cinaruco in Venezuela. The speckled peacock bass has been stocked in reservoirs constructed on blackwater rivers, such as Guri Reservoir on the Caroni River in Venezuela, and Balbina Reservoir on the Uatumã River in Brazil. The species also was stocked in reservoirs constructed on clearwater and whitewater rivers in Venezuela where it supports thriving sport fisheries that are easily accessed by people from large urban areas.

Geographic distribution of *Cichla temensis*.

Research conducted on the Cinaruco River in the Venezuelan Llanos revealed differences in the kinds of habitat used by of *C. temensis* and two other peacock bass species (*C. intermedia* and *C. orinocensis*) [5]. *Cichla temensis* and *C. orinocensis* are both common in channels and floodplain lakes, but only *C. orinocensis* typically occurs in small creeks. In river channels, the two species generally are found near shore in areas with submerged sticks and logs lacking strong current. In these nearshore habitats, the speckled peacock bass typically is found in deeper water further from the shoreline where it waits for prey to venture away from cover near shore. *Cichla temensis* sometimes occurs along sandbanks (point bars on the inside bend of river meanders) where the water has created series of ridges and troughs. In these habitats, speckled peacock bass often rest in troughs and attack prey passing over. *Cichla intermedia* essentially is restricted to channel habitats with flowing water containing rocks or logs.

The Cinaruco River in Venezuela's Santos Luzardo National Park in the Llanos supports a large population of speckled peacock bass. *Photo: K. Winemiller.*

A large fish that was lurking within a trough in a sandbar in the Cinaruco River before it attacked a surface lure as it passed overhead. *Photo: K. Winemiller.*

In the Pasimoni, Casiquiare, and Rio Negro, *C. temensis* and *C. orinocensis* can be found together in the river channel and floodplain lakes. Speckled peacock bass tend to be captured less frequently from the main river channel in the Pasimoni when compared to captures in the Cinaruco. The two species also co-occur in the Ventuari River, which is very rocky in its lower reaches. Here both species are caught from rocky habitats in slack water or places where rocks form a barrier against the water current. In both habitats of the Ventuari, *Cichla temensis* was always found to occupy deeper areas than *C. orinocensis*. In tributaries of the lower Rio Branco, such as the moderately blackwater Xeriuini and Tapera rivers, *Cichla temensis* co-occurs with *C. orinocensis*, *C. ocellaris* var. *monoculus*, and other kinds of cichlids, such as *Crenicichla reticulata*, *Hoplarchus psittacus*, *Heros notatus*, *Satanoperca lilith*, and *Hypselecara coryphaenoides* [6].

Rocky shoals of the Ventuari River in Venezuela provide excellent habitat for large speckled peacock bass. *Photo: C. Montaña.*

Three species of *Cichla* caught from the same location in the Casiquiare River: *C. orinocensis* (top), *C. ocellaris* var. *monoculus* (middle), and *C. temensis* (bottom). *Photo: C. Montaña.*

Cichla temensis, like *C. orinocensis*, is not endemic to the Caroní River, a blackwater river draining the Guiana Shield in Venezuela, but the two species were stocked into Guri Reservoir in 1979. Guri Reservoir is formed by the largest hydroelectric dam in the Orinoco Basin, and it is unusual in having extremely acidic black water. Other Venezuelan reservoirs constructed for hydropower or irrigation, such as Las Majaguas, Guárico, Masparro, and Coromoto, have clear water with relatively neutral pH, but some have been degraded by pollution from agricultural and municipal sources. *Cichla temensis* does not thrive as well as *C. orinocensis* in those reservoirs.

Because of its large size and superior qualities as a sport fish, the speckled peacock bass has been introduced into lakes and canals in several tropical and subtropical regions well beyond the Amazon and Orinoco basins. We have heard reports of *C. temensis* being introduced into the Brokopondo Reservoir in Suriname, and some tropical importers even sell juvenile *C. temensis* claimed to originate from Brokopondo. We have not yet been able to verify this.

Carlos Lasso Alcala of the Instituto Alexander von Humboldt in Colombia recently reported that *C. temensis* are being caught by anglers from reservoirs within the Magdalena River Basin in Colombia. Two other peacock bass species (*C. ocellaris* and *C. orinocensis*) also have been documented in lakes of this basin which lacks any native peacock bass. Surely, it is only a matter of time before these peacock bass escape into natural habitats.

In the 1970s, *Cichla temensis* and *C. ocellaris* were introduced by the state fishery agency into canals in and around Miami, Florida, but only the latter species became established [7]. It appears that the winter temperatures and/or neutral water chemistry of the region are unsuitable for *C. temensis*, which generally is found in rivers with pH less than 6 and sometimes as low as 4. During the 1970s and 1980s, the Texas fishery agency stocked thousands of *C. temensis* into several reservoirs receiving heated effluent from power plants, but the fish were unable to tolerate cool temperatures during winter [8], not to mention unsuitable water chemistry. Research on temperature tolerance indicated temperatures of 16°C (61°F) are lethal for this species [7]. Few survived more than a year, all perished within a few years, and the idea was abandoned. In 2011 three *C. temensis* were caught by anglers from the Upper Seletar Reservoir in Singapore [9]. These fish measured only 16–20 cm and likely were introduced by aquarists. It is unknown if the species has established a breeding population.

Recently, fisheries biologists proposed introducing *C. temensis* into reservoirs of Puerto Rico where *C. ocellaris* already supports sport fisheries [10]. The rationale for stocking speckled peacock bass is that their larger size would allow them to control populations of introduced tilapia

A speckled peacock bass lurking amid submerged logs in a clearwater tributary of the lower Rio Negro. *Photo: K. Winemiller.*

(African cichlids) more effectively than could *C. ocellaris*. However, *C. temensis* likely will not tolerate the neutral water chemistry of reservoirs fed by runoff from Puerto Rico's mountain ranges.

Feeding

Analysis of speckled peacock stomach contents revealed that specimens from the Cinaruco and Pasimoni rivers had fed entirely on fish [11]. At least 63 species from 22 families were recovered from stomachs of *C. temensis* from the Cinaruco River. Fish from the La Guardia River, a smaller river in the same region of the Llanos, had consumed 46 fish species from 14 families [12]. A study of *C. temensis* from the Ventuari River in the rainforest region of Venezuela's Amazonas state showed their diet was comprised of 80% fish, 10% shrimp, and 10% other items [13]. Analysis of gut contents from speckled peacock bass from the Ventuari River revealed at least 46 fish species from 16 families [13]. In all three rivers, the diet of *C. temensis* was dominated by tetras and their relatives (fish belonging to the order Characiformes), followed by catfish (order Siluriformes) and cichlids. The diet of *C. temensis* in Guri Reservoir is comprised entirely of fish, with the tetras *Hemigrammus micropterus* and *Bryconops caudomaculatus* most commonly consumed [14, 15]. Juvenile *C. temensis* in Brazilian reservoirs were found to consume zooplankton [16].

Two tetras (*Moenkhausia* species) recovered from the stomach of a Cinaruco speckled peacock bass. *Photo: C. Montaña.*

A positive relationship between predator length and prey length was reported for *C. temensis* in the Aguaro [17] and Cinaruco [18] rivers as well as Guri Reservoir [15]. In the Pasimoni River, large *C. temensis* (60–80 cm) have been observed attacking schools of relatively large fish, including bocachicos (*Semaprochilodus kneri*) and pike characins (*Acestrorhynchus* species), in shallow water near shore [11].

A healthy *C. temensis* from the Pasimoni River showing the three-barred pattern. *Photo: K. Winemiller.*

Seasonal variation in the water level of tropical rivers strongly influences ecological dynamics, including the feeding behavior, growth, and reproduction of peacock bass. As water levels fall during the dry season, prey densities increase in shrinking aquatic habitats, which results in more frequent interactions between predators and prey [19]. Large *C. temensis* in the Cinaruco River have been observed attacking schools of migrating bocachicos (*Semaprochilodus kneri*) during the early phase of the annual falling-water period [5]. These bottom-feeding fish dominate the diet of speckled peacock bass larger than 40 cm but are rare or absent in diets of smaller fish. *Cichla orinocensis* and *C. intermedia* in the Cinaruco are too small to consume this seasonally abundant prey. During the falling-water period, young-of-the-year bocachicos migrating into the Cinaruco River from the productive Orinoco floodplains and provide a major nutritional subsidy for resident *C. temensis*. Researchers estimate that consumption of these migratory fish contributes significantly to improved body condition of speckled peacock bass prior to their spawning season [20]. Just before the bocachicos begin their ascent of the Cinaruco River, many speckled peacock bass move from floodplain lakes into the main channel.

The bocachico, *Semaprochilodus kneri*, a migratory fish from the Cinaruco River. *Photo: K. Winemiller.*

Seasonal shifts in prey availability affect peacock bass feeding. For example, in a study conducted in three rivers in Venezuela (Cinaruco, La Guardia, Ventuari), *C. temensis* consumed larger prey during the falling-water period than during the late dry season when aquatic habitat is reduced to is annual minimum [12]. As large prey become less abundant toward the end of the dry season, as a consequence of predation mortality, peacock bass consume progressively smaller prey.

Cinaruco *C. temensis* showing the three-barred pattern that is typical of the dry season. *Photo: C. Montaña.*

The speckled peacock bass ranks among the largest and most voracious predators in South American rivers, but, paradoxically, it feeds at a relatively low trophic level [21]. *C. temensis* biomass in the Cinaruco River is largely supported by detritivorous bocachico and omnivorous fish, such as *Moenkhausia*, *Metynnis*, and *Myleus* species.

Cannibalism and predation mortality

At least two studies have documented cannibalism by *C. temensis* in the Cinaruco River [5, 12]. Cannibalism by peacock bass also was reported for fish in the La Guardia and Ventuari rivers in Venezuela [12].

Animals that prey upon speckled peacock bass include freshwater dolphins (*Inia geoffrensis*), river otters (*Pteronura brasiliensis*), and caiman (*Caiman crocodilus, Melanosuchus niger*). River dolphins are particularly adept at capturing fatigued fish released by anglers. Piranhas usually bite only the fins of peacock bass, but they can cause more serious damage to fish that are fighting on the end of an angler's line. Juvenile peacock bass are vulnerable to predation by a great diversity of fish and birds.

This small peacock bass was quickly devoured by hungry piranhas before it could be brought to the boat. *Photo: K. Winemiller.*

This speckled peacock bass was attacked by a river dolphin while being retrieved by the angler. *Photo: C. Montaña.*

Growth

In Venezuelan rivers, speckled peacock bass are larger and grow faster than *C. orinocensis* and *C. intermedia* [18]. Based on analysis of otolith growth rings and data pooled from multiple Venezuelan stocks, the average growth was 4.6 cm per year. The maximum length was 81.5 cm and the maximum age was 9 years, both recorded from fish captured from the Pasimoni River [11, 18].

Growth of *C. temensis* in the Cinaruco during the first year of life averaged 2.7 cm per month [18]. Analysis of growth rings on scales from *C. temensis* obtained from the middle Rio Negro and floodplain lakes estimated the length of sexual maturation at 31 cm, maximum length at 68 cm, and longevity at 14 years [22]. *Cichla temensis* reared in Texas ponds grew between 1.3 and 6.1 cm per month during the first 6 months after hatching [23], suggesting that fish reared in captivity may grow faster than fish in natural habitats.

Based on data from multiple *C. temensis* stocks in Venezuela, the average minimum size of maturation is 36 cm for males and 32.5 cm for females [13, 18]. Captive *C. temensis* attained sexual maturity at 34 cm [24]. The length-weight relationship varies little among *C. temensis* stocks from various rivers and reservoirs studied in Venezuela [3, 11].

This small fish is emaciated and probably heavily parasitized, and yet it shows a vivid coloration pattern. *Photo: C. Montaña.*

Cichla temensis is the largest cichlid in South America and perhaps even in the world, with only *Boulengerochromis microlepis*, the giant cichlid from Lake Tanganyika in Africa, competing for the title (the IGFA all-tackle record for the giant cichlid is only 2.94 kg; Fishing World Records.com lists a fish at 5 kg measuring 80 cm total length). The IGFA world record for the speckled peacock bass is listed at 13.19 kg (29 lbs, 10 oz) and measured 90 cm total length. This huge fish was caught by Andrea Zaccherini near Santo Isabel do Rio Negro, Brazil. In December 2017, an even larger fish (14 kg [30 lbs, 13 oz], measuring 1.08 m in total length) was caught by angler Marcel Griot in the Marié River, a blackwater tributary of the upper Rio Negro in Brazil, but that record is not currently listed by the IGFA. Some people have claimed to have seen *C. temensis* weighing 16 kg (over 35 lbs), but this has never been documented.

Population abundance and structure

Best approximations of population density and size structure of speckled peacock bass in rivers are from studies conducted on the Cinaruco River, a clearwater tributary of the Orinoco in the Venezuelan Llanos. Males were captured more often than females during angling surveys [3, 18]. Overall, fish ranged from 16 to 72 cm and 0.1 to 6.8 kg. Males averaged 41 cm and 1.5 kg, and females averaged 40 cm and 1.6 kg [18].

Speckled peacock bass showing the three-barred pattern. This large fish was caught in the Rio Negro by Clint Robertson. *Photo: K. Winemiller.*

In the Ventuari River, a tributary of the upper Orinoco, *C. temensis* averaged 40cm and 1.5kg [13]. *Cichla temensis* captured from the Pasimoni and other rivers of the Casiquiare region in southern Venezuela measured from 30 to 81.5cm and 0.5 to 9.2kg [18]. Males averaged 60cm and 4.5kg, and females averaged 59cm and 4.9kg.

Based on a gillnet survey conducted in Venezuela's Guri Reservoir during 1993–94, speckled peacock bass measured from 6.5 to 70cm and weighed up to 5.4kg [15]. A small sample of *C. temensis* from Venezuela's Las Majaguas Reservoir ranged in size from 34 to 52cm with an average weight of 1.9kg.

During 1999–2003, we and our colleagues tagged 1100 *C. temensis* during a mark-recapture study of speckled peacock bass in the Cinaruco River [25]. The recapture rate was low (3.8%), an indication that the population is very large, the mortality rate was high, or the plastic tags were being lost, perhaps to fin-nipping piranhas (*Serrasalmus* species). Most recaptures were within 1km of the tagging location, but some fish moved over 20km. Larger fish tended to travel longer distances. During the annual period of falling water levels, many fish shifted from locations in the river channel into floodplain lakes. At that time, anglers who had visited this river over several years reported recent reductions in the maximum size of speckled peacock bass. Catch rates for speckled peacock bass were high, but impressively large fish were rare. In the years following our mark-recapture study, illegal fishing activity increased on the Cinaruco River, and the current status of the river's peacock bass stocks is unknown due to the lack of research because of the unfortunate socioeconomic and security situation in the country.

Fish with numbered plastic "spaghetti" tag attached for a mark-recapture study. *Photo: C. Montaña.*

A study of *C. temensis* movement in the middle Rio Negro used otolith microchemistry to determine changes in locations over the lifespan of individual fish [26]. This method takes advantage of regional differences in geology and water chemistry that are recorded in the growth layers of a fish's otoliths. That study similarly concluded that some fish (42% of the sample) were collected from the same area where they originated as juveniles, whereas other fish had otolith chemical patterns indicating movement between regions of the middle Rio Negro.

Speckled peacock bass from a lake in the floodplain of the lower Branco River. *Photo: K. Winemiller.*

The Aguaro River in the Central Llanos region of Venezuela once contained a thriving population of speckled peacock bass, but during the 1980s and 1990s, illegal fishing with large seine nets was rampant and depleted the stock [27]. The most recent assessments found the speckled peacock bass to be extremely rare, but the butterfly peacock bass was still fairly common but with a reduced average size of fish in the population [18, 27].

A study of speckled peacock genetics revealed relatively low levels of genetic divergence among populations from different regions within the geographic range of the species [28]. The genetic analysis confirmed that stocks from various rivers interbreed naturally and therefore belong to a single species. However, the study also inferred that modest genetic differences could be associated with adaptation to local conditions, and therefore local populations should be managed as separate stocks. Another study of *C. temensis* genetic variation across its geographic distribution concluded that whitewater rivers (i.e., neutral pH with high sediment loads causing low transparency) pose barriers to dispersal by this species [29].

Speckled peacock bass in the fish market of Manaus, Brazil. *Photo: L. Kelso Winemiller.*

Reproduction

Most studies report evidence of spawning by at least some peacock bass during almost any time of the year. However, spawning usually peaks near the end of the falling-water phase of the annual flood pulse (water levels usually, but not always, fall during the dry season), with nesting and brood guarding lasting into the early part of the annual rising water phase (which usually coincides with the early rainy season) [3, 18, 22, 30]. The coloration pattern gradually changes from the speckled pattern to the three-barred pattern as gonads develop, with no apparent coloration difference between the sexes [3, 4]. Prior to spawning, male *C. temensis* become aggressive and develop a nuchal hump, a bulge created by deposition of fat on top of the head. The nuchal hump of cichlids is a secondary sexual character that visually communicates to females the male's robust physical condition and suitability as a mate [31]. Some biologists speculate that this fatty deposit may provide additional energy reserves to sustain males during the period of spawning and nest guarding. Both sexes acquire more orange or orange-red coloration on the branchiostegal membranes, chest, and fins [32].

In the Cinaruco River, most adult *Cichla temensis* transition from the speckled coloration pattern to the three-barred pattern from December to February, and spawning commences during March with peak activity during April–May [3]. On numerous occasions, we have observed nesting pairs of *C. temensis* in shallow water near the shoreline of lakes in the floodplain of the Cinaruco River. These nests were shallow pits in the sand where leaf litter and sticks had been swept away. Nests of *C. temensis* have been observed in the Ventuari River on sand banks and rocky substrates at depths less than 50cm [13]. In the Casiquiare and upper Orinoco region, nests and males with nuchal humps often were observed during expeditions made during January and February.

Cichla temensis captured from a lake in the floodplain of the Cinaruco River. This male specimen shows the three-barred pattern and nuchal hump indicative of the nesting season. *Photo: K. Winemiller.*

Constricted growth of otoliths at the end of the low-water period and start of the annual rains indicates lower food intake by Cinaruco speckled peacock bass during reproductive activity [18]. Findings from stomach contents analysis partially confirm this hypothesis. For example, *C. temensis* had the highest proportion of empty stomachs during the end of the low-water and beginning of rising-water period, which coincides with gonad maturation. Similar findings were reported for the Aguaro, La Guardia, and Ventuari rivers in Venezuela [5, 12, 13]. In the Rio Negro, speckled peacock bass fat stores and growth are reduced during the reproductive period [22]. Spawning by *C. temensis* is more synchronized in rivers that undergo strong seasonal changes in hydrology, such as those in the Llanos region, when compared to populations in rainforest rivers of the Upper Orinoco in Venezuela and Rio Negro in Brazil and those in reservoirs [3].

The mean number of mature eggs per female (batch fecundity) of *C. temensis* from the Cinaruco River was 5725 [5, 11]. For Ventuari *C. temensis*, mean fecundity was documented as 5200 eggs [13].

Speckled peacock bass nest in a sandbank of the Ventuari River. The eggs are attached to a stick. After hatching, the fry may remain in the nest for several days while being guarded by both parents. The parents continue guarding the free-swimming fry as they forage for rotifers and other microscopic organisms. *Photo: C. Montaña.*

Dave Jepsen with a beautiful male *C. temensis* from the Pasimoni River with a well-developed nuchal hump and showing the three-barred pattern characteristic of fish during the spawning season. *Photo: K. Winemiller.*

References

[1] K.O. Winemiller, H. López Fernández, D.C. Taphorn, L.G. Nico, A. Barbarino Duque, Fish assemblages of the Casiquiare River, a corridor and zoogeographic filter for dispersal between the Orinoco and Amazon basins, J. Biogeogr. 35 (2008) 1551–1563.
[2] K.O. Winemiller, S.C. Willis, Biogeography of the Casiquiare River and Vaupes Arch, in: J.S. Albert, R.E. Reis (Eds.), Historical Biogeography of Neotropical Fishes, Verlag Dr. Friedrich Pfeil Scientific Publishers, Munich, Germany, 2011, pp. 225–242 (Chapter 11).
[3] K.O. Winemiller, Ecology of peacock cichlids (*Cichla* spp.) in Venezuela, J. Aquaric. Aquat. Sci. 9 (2001) 93–112.
[4] P. Reiss, K. Able, M. Nunes, T. Hrbek, Color pattern variation in *Cichla temensis* (Perciformes: Cichlidae): resolution based on morphological, molecular, and reproductive data, Neotrop. Ichthyol. 10 (2012) 59–70.
[5] D.B. Jepsen, K.O. Winemiller, D.C. Taphorn, Temporal patterns of resource partitioning among *Cichla* species in a Venezuelan blackwater river, J. Fish Biol. 51 (1997) 1085–1108.
[6] E. Ferreria, J. Zuanon, B. Forsberg, M. Goulding, S.R. Briglia-Ferreira, Rio Branco—Peixes, ecologia e conservação de Roraima, Amazon Conservation Association (ACA), 2007, p. 201.
[7] W.C. Guest, B.W. Lyons, G. Garza, Effects of temperature on survival of peacock bass fingerlings, in: Proceedings of the Annual Conference of the Southeastern Association of Fish and Wildlife Agencies, vol. 33, 1980, pp. 620–627.
[8] R.G. Howells, G.P. Garrett, Status of some exotic sport fishes in Texas waters, Tex. J. Sci. 44 (3) (1992) 317–324.
[9] J.H. Liew, H.T. Heok, D.C.J. Yeo, Some cichlid fishes recorded in Singapore, Nat. Singap. 5 (2012) 229–236.
[10] J.W. Neal, J.M. Bies, C.N. Fox, C.G. Lilyestrom, Evaluation of proposed speckled peacock bass *Cichla temensis* Introduction to Puerto Rico, N. Am. J. Fish Manag. 37 (5) (2017) 1093–1106.
[11] K.O. Winemiller, D.C. Taphorn, A. Barbarino Duque, The ecology of *Cichla* (Cichlidae) in two blackwater rivers of southern Venezuela, Copeia 1997 (1997) 690–696.
[12] C.G. Montaña, C.A. Layman, K.O. Winemiller, Gape size influences seasonal patterns of piscivore diets in three Neotropical rivers, Neotrop. Ichthyol. 9 (2011) 647–655.
[13] C.G. Montaña, D.C. Taphorn, C.A. Layman, C. Lasso, Distribución, alimentación and reproducción de tres species de pavones *Cichla* spp. (Perciformes, Cichlidae) en la cuenca del Rio Ventuari, Venezuela, Mem. Fund. La Salle Cienc. Nat. 165 (2007) 83–102.
[14] D. Novoa, Aspectos generales sobre la ecologia, pesqueria, manejo y cultivo del pavon (*Cichla orinocensis* and *Cichla temensis*) en el Lago Guri y otras areas de la region Guyana, Natura 96 (1993).
[15] J.D. Williams, K.O. Winemiller, D.C. Taphorn, L. Balbas, Ecology and status of piscivores in Guri, an oligotrophic tropical reservoir, N. Am. J. Fish Manag. 18 (1998) 274–285.
[16] R. Braga, Crecimiento de tucunare pinima, *Cichla* Humboldt, em cativiero (Actinopterygii, Cichlidae), Dusenia 4 (1953) 41–47.
[17] D. Rodríguez-Olarte, D.C. Taphorn, Ecologia trofica de *Cichla orinocensis* Humbolt 1833 (pisces, Teleostei, CIchlidae) en un humedal de los llanos centrales de Venezuela, Biollania 13 (1997) 139–163.
[18] D.B. Jepsen, K.O. Winemiller, D.C. Taphorn, D. Rodriguez Olarte, Age structure and growth of peacock cichlids from rivers and reservoirs of Venezuela, J. Fish Biol. 55 (1999) 433–450.
[19] C.A. Layman, K.O. Winemiller, Size-based response of prey to piscivore exclusion in a species-rich Neotropical river, Ecology 85 (2004) 1311–1320.
[20] D.J. Hoeinghaus, K.O. Winemiller, C.A. Layman, D.A. Arrington, D.B. Jepsen, Effects of seasonality and migratory prey on body condition of *Cichla* species in a tropical floodplain river, Ecol. Freshw. Fish 15 (4) (2006) 398–407.
[21] C.A. Layman, K.O. Winemiller, Patterns of habitat segregation among large fishes in a Neotropical floodplain river, Neotrop. Ichthyol. 3 (2005) 111–117.
[22] C.P. Campos, C.E. de Carvalho Freitas, S. Amadio, Growth of the *Cichla temensis* Humboldt, 1821 (Perciformes: Cichlidae) from the middle Rio Negro, Amazonas, Brazil, Neotrop. Ichthyol. 13 (2) (2015) 413–420.
[23] W. Rutledge, B. Lyons, Texas peacock bass and Nile perch: status report, in: Proceedings of the Annual Conference Southeastern Association of Fish and Wildlife Agencies, vol. 39, 1976, pp. 18–23.
[24] W.S. Devick, Life history study of the tucunaré *Cichla ocellaris*, in: Federal Aid in Sportfish Restoration Project F-9-1, Job Completion Report, Hawaii Department of Land and Natural Resources, Honolulu, 1972.
[25] D.J. Hoeinghaus, C.A. Layman, D.A. Arrington, K.O. Winemiller, Movement of *Cichla* species (Cichlidae) in a Venezuelan floodplain river, Neotrop. Ichthyol. 1 (2003) 121–126.

[26] R.G.C. Sousa, R. Humston, C.E.C. Freitas, Movement patterns of adult peacock bass *Cichla temensis* between tributaries of the middle Negro River basin (Amazonas—Brazil): an otolith geochemical analysis, Fish. Manag. Ecol. 23 (2016) 76–87.

[27] D. Rodriguez-Olarte, D.C. Taphorn, Distribución y abundancia de pavones (*Cichla orinocensis* y *C. temensis*, Pisces:Cichlidae) en un humedal de los Llanos Centrales en Venezuela, Vida Silv. Neotrop. 8 (1–2) (1999) 43–50.

[28] S.C. Willis, K.O. Winemiller, C.G. Montaña, J. Macrander, P. Reiss, I.P. Farias, G. Ortí, Population genetics of the speckled peacock bass (*Cichla temensis*), South America's most important inland sport fishery, Conserv. Genet. 16 (2015) 1345–1357.

[29] J. Macrander, S.C. Willis, S. Gibson, G. Orti, T. Hrbek, Polymorphic microsatellite loci for the Amazonian Peacock Basses, *Cichla orinocensis* and *C. temensis*, and cross-species amplification in other *Cichla* species, Mol. Ecol. Resour. (2012), https://doi.org/10.1111/j.1755-0998.2012.03173.x.

[30] R.H. Lowe-McConnell, The cichlid fishes of Guyana, South America, with notes on their ecology and breeding behavior, Zool. J. Linn. Soc. 48 (1969) 255–302.

[31] G. Barlow, The Cichlid Fishes: Nature's Grand Experiment in Evolution, Perseus Publishing, Cambridge, MA, 2000.

[32] O. Fontenele, Contribução para o conhecimiento de biologia dos tucanarés, Actinopterygii Cichlidae, em cativiero. Aprelhlo de reproducao. Habitos de desova incubaçã, Rev. Bras. Biol. 10 (1950) 503–519.

CHAPTER

6

Pinima peacock bass, *Cichla pinima* (Kullander & Ferreira 2006)

Cichla pinima

Origin of name

Tupi-Guarana adjective "*pinima*" meaning "spotted with white"

Putative synonyms

Cichla jariina, *C. thyrorus*, *C. vazzoleri*; Kullander & Ferreira 2006

Common names

Pinima peacock bass, Yellow peacock bass, Tucunaré açu (a name also applied to *C. temensis*), Tucunaré pinima

Geographic distribution

Cichla pinima inhabits clearwater rivers draining the Guiana and Brazilian shield regions of the lower Amazon (eastern Amazon). It is largely restricted to lowland rivers with broad floodplains, including floodplains of the lower Amazon River, from near the mouth of the Madeira River to the estuary and adjacent coastal drainages.

Natural history

Found in lakes, floodplain creeks, and river channels in areas with no or slow current, usually near shore. Primarily piscivorous.

https://doi.org/10.1016/B978-0-323-85157-2.00009-3

Several years ago, we (KOW, LKW) reared our first two peacock bass in our home aquaria. Upon arrival of the shipment from the importer, the juvenile *C. temensis* were placed in a small aquarium until they reached a size sufficient for them to avoid becoming prey of the arowana in our 300-gal aquarium. The little peacocks were offered all the mosquitofish they cared to devour, and consequently they grew rapidly. Actually, their growth rate was quite impressive, though we should have anticipated this given the fact that *C. temensis* is the largest peacock bass species. However, as the fish continued to grow, and their coloration pattern gradually shifted from the juvenile pattern (solid horizontal stripe along the head and body) to the subadult pattern (disappearance of the horizontal stripe, formation of lateral blotches and caudal ocellus, and development of a yellow background coloration), it became increasingly apparent that these were not *C. temensis*. In fact, their coloration pattern did not match that of any peacock species recognized by ichthyologists at that time.

With the publication of K&F's monograph on the taxonomy of *Cichla* [1], we soon realized that our peacock bass were indeed *C. pinima*. The coloration pattern matched the description, and the supplier's misidentification of the juveniles is completely understandable given that *C. temensis* and *C. pinima* are nearly indistinguishable at a small size. Therefore we didn't have the world's largest species of peacock bass but actually had something more unique—a newly described species.

We enjoyed observing the two fish grow and interact with one another and their tankmates. In addition to the arowana, there was a parrot cichlid (*Hoplarchus psittacus*), two severums (*Heros severus*), and a tiger shovelnose catfish (*Pseudoplatystoma tigrinum*). Mosquitofish and minnows were provided as prey. Over the years, we've kept a variety of predatory fish in aquaria, but the peacock bass ranks at the top of our list as the most aggressive in terms of speed and efficiency of attack. When hungry, these fish pursue prey like a heat-seeking guided missile. Another type of predatory fish that approaches the level of aggression of the peacock bass attack is the Asian snakehead. When I (KW) was in graduate school, my roommate claimed that the snakehead in my aquarium gave him nightmares after watching it feed. Interestingly, as voracious and fierce as peacock bass are when preying on small fish, they are surprisingly calm with larger tankmates. In general, South American cichlids are not as pugnacious as many of the Central American and African cichlids. Our parrot cichlid and severum cichlids acted more aggressively toward the peacock cichlids than did the peacocks toward their tankmates, that is provided the latter were sufficiently large to avoid become a meal.

Large predatory fish make interesting pets but require large aquaria, and therefore should only be kept by specialist hobbyists. Large fish produce considerable amounts of waste, and consequently their aquarium requires adequate filtration. In addition, peacock bass are piscivores and should be fed fish. It can be difficult and expensive to obtain a steady supply of live fish, but we were fortunate to have a garden pond full of mosquitofish supplemented by minnows purchased at the local bait shop. Depending on the species, peacock bass can live more than 10 years, thus caring for captive fish is a long-term commitment. In our case, the peacock bass eventually grew to a size that we considered too large even for our 300-gal tank. We eventually donated one of them for display at a public aquarium, and the other one enjoyed a long Texas summer in the large pool within our garden pond. The plan was to remove the fish before temperatures got too cold in the fall, but unfortunately, we miscalculated, and on a chilly night the fish succumbed. That specimen now resides in the ichthyology teaching collection at our university. Peacock bass are tropical fish, and experiments conducted by fisheries management agencies in Florida and Texas determined that peacock bass do not tolerate temperatures below about 15°C (59°F). Florida fisheries managers successfully established *Cichla ocellaris* in canals of south Florida, but efforts to establish peacock bass in Texas reservoirs receiving heated water from power plants ultimately failed. As result of these research findings and failed attempts to establish stocks in Texas, the state agency lifted the ban on import and possession of live peacock bass.

When this fish was sold to the authors, it was claimed to be a juvenile *C. temensis*. It eventually turned out to be *C. pinima* and lived for several years in the authors' 300-gal aquarium and later was moved to an outdoor pond.

Identification

Described fairly recently, *Cichla pinima* is similar in appearance, size, and ecology to *C. temensis*, and juveniles of the two species are nearly indistinguishable. According to K&F, the lateral line is always continuous in *C. temensis* and discontinuous (meaning with a gap between an upper anterior portion and a lower posterior portion) in *C. pinima*. Both of these species generally display some form of vertical bars in positions 1, 2, and 3, but *Cichla pinima* is distinguished from *C. temensis* by having a large ocellated black spot below the rayed dorsal fin (approximately in position of vertical bar 3). The vertical bars in *C. pinima* are often broken into a large spot near the midline, and blotches or bars above and below, all of which may be bordered by white or yellow spots to give an appearance of ocellation. The spot at the midline of the first bar is often much larger than the portions of bar 1 located above and below the spot. *Cichla pinima* also has much less extensive blotching and spotting on the postorbital and opercular areas of the head.

Cichla pinima caught from the Maguari River close to the city of Belém in the estuarine region of the Amazon. This specimen shows extensive white spots. *Photo: Leonardo Okada.*

The background coloration of the head and body of subadults and adults varies from yellow to olive-green to green, with the lower portion of the head and body lighter and the dorsal region darker. In breeding individuals, especially males, the vertical bars become darker and more distinct, with dorsal portions of each bar often detached as blotches that are ocellated. Some fish, both breeding and nonbreeding, have numerous white spots irregularly distributed all over the head and sides of the body. In some individuals, the white spots on the head are joined to form a reticulated (wormy) pattern. Prior to and during the reproductive period, the branchiostegal membranes, throat, and belly are yellow or yellow-orange (in *C. temensis*, these areas are orange, orange-red, or red). The pelvic, anal, and lower half of the caudal fin are yellow, olive green, green, or blue-green in nonbreeding adults. The dorsal fin and top half of the caudal fin are gray or blue-gray and densely covered in small, translucent white spots.

Small juveniles have a continuous dark lateral stripe that extends from the eye to the base of the caudal fin. In larger juveniles, three dark bars intersect the horizontal stripe at bar positions 1, 2, and 3. As juveniles grow into subadults, the horizontal stripe disappears and the caudal ocellus becomes distinct. The background coloration is gray or light brown and covered by small white spots. The fins are clear or transparent gray and become gray with transparent white spots as the fish grow.

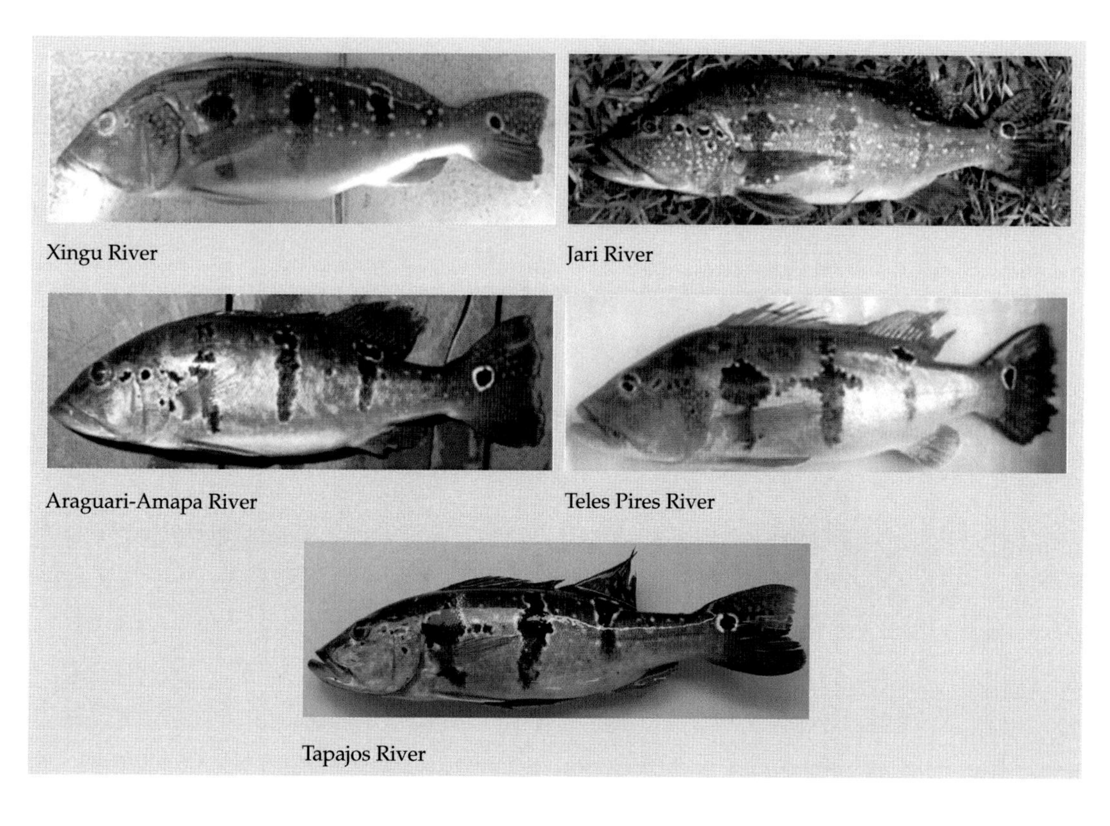

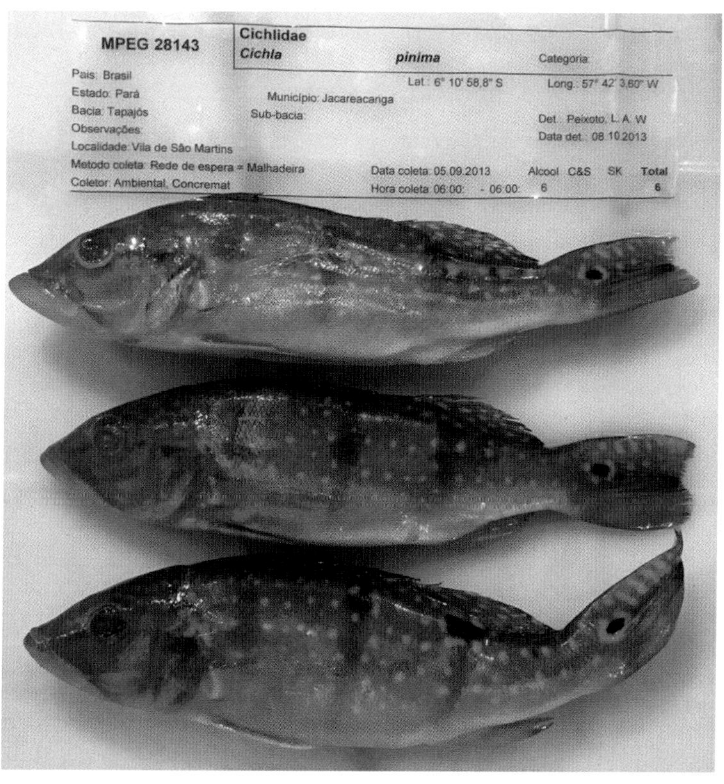

Preserved juvenile specimens of *C. pinima* archived in the Ichthyology Collection of the Museu Emilio Goeldi in Belém, Brazil. *Photo: K. Winemiller.*

Preserved specimens of juvenile *C. pinima* in the Ichthyology Collection of the Museu Emilio Goeldi. These specimens were collected from a region of Amazon estuary and misidentified as *C. temensis*. *Photo: K. Winemiller.*

Subadult *C. pinima* from the Tapajos River near Alter do Chau, Brazil. *Photo: S. Willis.*

Distribution and habitat

Ecological information provided by K&F for *C. pinima* (and the three species considered synonymous by Willis et al. [2–4]) is limited to description of the geographic distribution. No other published studies on the ecology of peacock bass stocks deal with populations that can be definitively identified as *C. pinima*.

Based on our current knowledge of the taxonomy and distribution of *C. pinima*, the species inhabits clearwater rivers draining the Guiana and Brazilian shield regions of the lower Amazon (eastern Amazon). This species generally is not found in upper reaches of major tributaries of the lower Amazon that have cataracts, either because these cataracts pose physical barriers to their upstream dispersal, or else this species doesn't survive or compete well in rocky habitats with fast-flowing water. The pinima peacock bass is largely restricted to lowland rivers with broad floodplains, including floodplains of the lower Amazon River, from near the mouth of the Madeira River to the estuary and adjacent coastal drainages. *Cichla pinima* is common in clearwater habitats of the lower Madeira where it sometimes comes into contact with *C. temensis*, the species to which it is most closely related. The two species apparently interbreed to some extent in these contact zones [4]. The geographic distribution of *C. pinima* does not overlap with the other four species in the clade A lineage of *Cichla* identified by Willis [4] (see Chapter 11).

In the Xingu River, the pinima peacock bass is restricted to the lowest reach, between the last rapids and the river's juncture with the Amazon River. The habitat in this region of the Xingu is markedly different than those found upstream where the channel is heavily braided and dominated by rapids. *Cichla melaniae* thrives within these faster flowing, rocky reaches. A similar

Pinima peacock bass from the Araguari River, tributary of the Madeira River. *Photo: S. Willis.*

Distribution of *C. pinima* in the Tapajos and eastern Amazon.

situation is observed in the Tocantins where *C. pinima* is found only in the lowermost reaches and is replaced by *C. piquiti* in areas above the rapids. In the Tapajos River, *C. pinima* ranges much further upstream to the lower parts of the Juruena and Teles Pires rivers. The rapids in the lower and middle reaches of this Brazilian Shield river apparently are not large enough to form a barrier to dispersal.

Within the Madeira Basin, *C. pinima* has an even larger distribution in the lower Madeira River, extending from the mouth upstream to the San Antonio Cataracts near the city of Porto Velho. These cataracts formed a major barrier to fish dispersal, but recently have been flooded by the Santo Antônio Dam. The species does not seem to occur anywhere in the Amazon River upstream from the mouth of the Madeira River.

Pinima peacock bass captured using a spear gun in the crystalline waters of the Lower Xingu River. *Photo: M. Sabaj.*

Cichla pinima from the Araguari-Amapa River. *Photo: S. Willis.*

In lower Amazon tributaries draining the Guiana Shield, the pinima peacock bass sometimes is found in headwater reaches, which must have resulted from historic dispersal past major rapids in middle and lower reaches. K&F described *C. jariina*, *C. thyrorus*, and *C. vazzoleri* from this region of the *C. pinima* distribution, but genetic analysis indicates these populations do not constitute distinct evolutionary lineages, and therefore should not be recognized as valid species [2–4]. It should be noted that here we are following the recommendation of Willis et al. [2–4] by designating *Cichla pinima* as the valid name for this taxon, rather than one of the other three names that were proposed by K&F in the same publication. This recommendation was given based on the much greater geographic distribution attributed to *C. pinima* relative to the other three named entities. At the same time, depending on the genetic dataset and analysis performed, there is evidence to support the existence of one (*C. pinima* sensu *lato*), two (a southern lineage and a western lineage), or four lineages (a southern lineage, western lineage, Aripuanã-Machado lineage, and Jatapu lineage). The southern and western lineages show clear evidence of admixture (results from interbreeding among individuals of the two lineages), and the Aripuanã-Machado and Jatapu lineages appear to be recently divergent

Preserved specimens archived in the Ichthyology Collection of the Instituto Nacional de Pesquisas do Amazonia (INPA) in Manaus. The two top specimens are from the Trombetas River drainage. K&F named the top specimen *Cichla vazzoleri* and the bottom specimen *C. thyrorus*. However, based on recent genetic evidence, these fish appear to differ little from *C. pinima*. The bottom fish was given the name *C. jariina* by K&F, and this fish also is not genetically distinct from *C. pinima* (see Chapter 11). *Photos: S. Willis.*

lineages. Future research that considers additional genes from more specimens from more locations using the latest methods of data analysis may ultimately reveal some of these lineages to be valid species. For now, it seems prudent to treat all of these populations as a single valid species, *C. pinima*. Nonetheless, regional stocks likely are adapted to their local environment and should be managed accordingly.

Cichla pinima often coexists with *C. ocellaris* var. *monoculus*, and apparently the two species use of habitat and foraging behavior differ in a manner consistent with *C. temensis* and *C. orinocensis*, the two species that co-occur within the Orinoco and Rio Negro basins. The pinima peacock bass grows larger and therefore is capable of feeding on larger prey than the butterfly peacock bass. The pinima peacock bass is assumed to inhabit deeper water located further from shore as compared with areas where butterfly peacock bass normally reside. Of course, both species exploit habitats in riparian areas during flood pulses, and both species are found in lakes, floodplain creeks, and river channels in areas with no or slow current, usually near shore.

Cichla pinima (top) and *C. ocellaris* var. *monoculus* (bottom) from a fish market on the lower Xingu River where these species coexist. *Photo: S. Willis.*

Cichla ocellaris var. *monoculus* (top) and *C. pinima* (bottom) from the fish market in the town of Itaituba on the bank of the Tapajos River. These two species are broadly distributed in the eastern Amazon Basin. *Photo: S. Willis.*

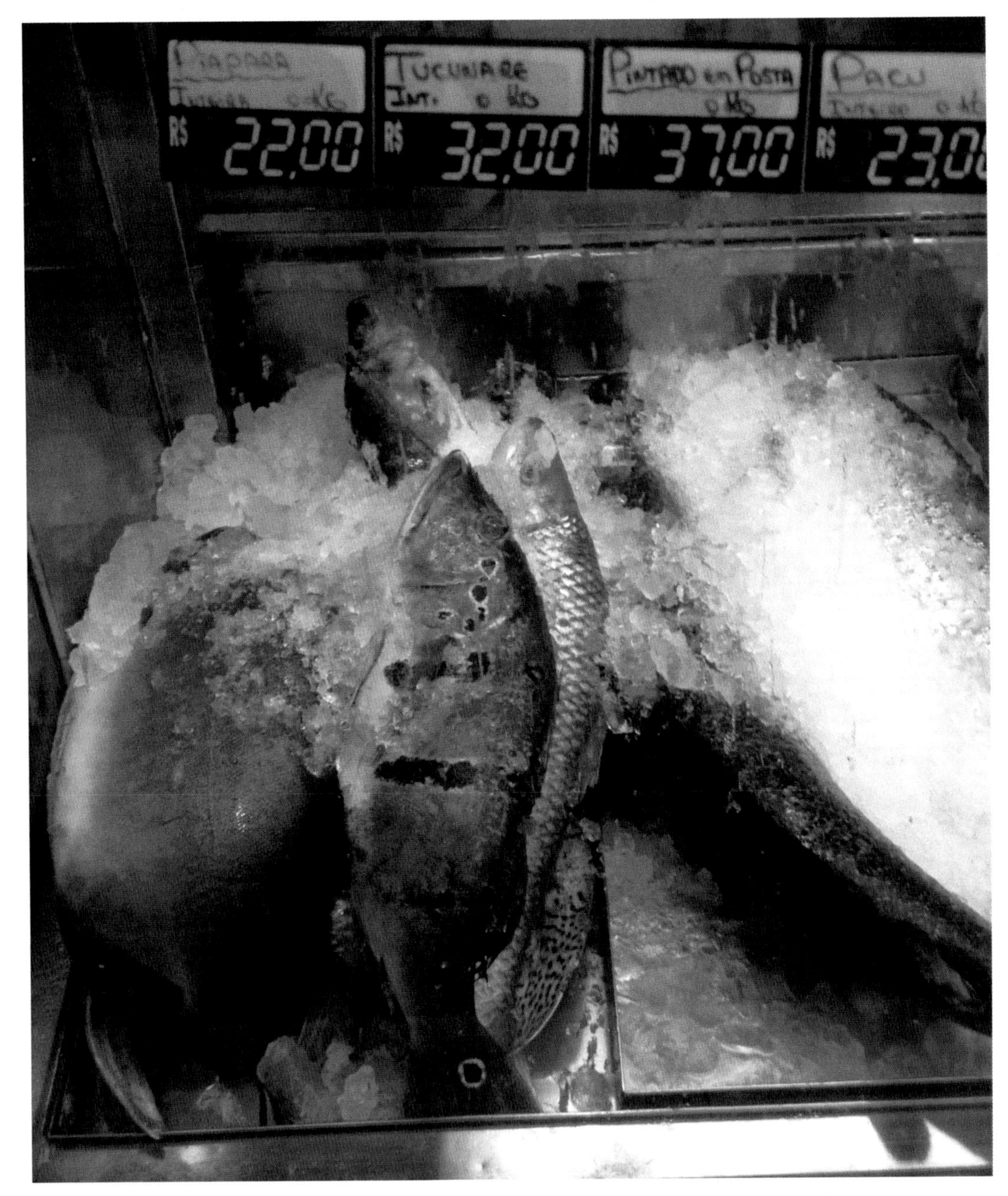

Pinima peacock bass for sale in a supermarket in Maringá, a city in Paraná state, southern Brazil. This location is far outside the geographic distribution of the species, and either was transported from a location in the upper Tapajos Basin or possibly was farmed at a closer location. *Photo: L. Kelso Winemiller.*

Feeding

Reports of feeding behavior or diet of *C. pinima* appear to be unavailable at this time.

Pinima peacock bass from the Teles Pires River, upper Tapajos Basin. *Photo: S. Willis.*

Growth

Reports documenting growth are lacking for *C. pinima* at this time. The IGFA world record reported for the pinima peacock bass is 11.09 kg (24 lbs, 7 oz), which ranks second largest among *Cichla* species. The location where this record fish was captured is listed as Barragem do Castanhão, which apparently is an artificial lake.

Preserved adult specimen archived in the Ichthyology Collection of the Museu Emilio Goeldi in Belém. *Photo: K. Winemiller.*

Population abundance and structure

Reports of population abundance and structure seem to be unavailable for *C. pinima* at this time. The species clearly is abundant in many regions within its native range, and important commercial and sport fisheries exploit this species.

Pinima peacock bass in the Manaus fish market. These fish may have come from clearwater tributaries of the lower Madeira River. *Photo: K. Winemiller.*

Reproduction

We have not been able to find any published accounts of the reproductive biology of *C. pinima*, which, again, is surprising given this species' large geographic distribution, large size, and high value in subsistence, commercial, and sport fisheries.

Male *C. pinima* from the Tapajos River showing a well-developed nuchal hump. *Photo: R. Silvano.*

Male pinima peacock bass with a large nuchal hump indicative of nesting or brood guarding. This fish was caught from the Maguari River in a location near the city of Belem that is under tidal influence. *Photo: L. Okada.*

Pinima peacock bass being released back into a pond. *Photo: K. Winemiller.*

References

[1] S.O. Kullander, E.J.G. Ferreira, A review of the South American cichlid genus *Cichla*, with descriptions of nine new species (Teleostei: Cichlidae), Ichthyol. Explor. Freshw. 17 (2006) 289–398.
[2] S.C. Willis, M.S. Nunes, C.G. Montana, I.P. Farias, N.R. Lovejoy, Systematics, biogeography, and evolution of the neotropical peacock basses *Cichla* (Perciformes: Cichlidae), Mol. Phylogenet. Evol. 44 (1) (2007) 291–307.
[3] S.C. Willis, J. Macrander, I.P. Farias, G. Orti, Simultaneous delimitation of species and quantification of interspecific hybridization in Amazonian peacock cichlids (genus *Cichla*) using multi-locus data, BMC Evol. Biol. 12 (2012) 96, https://doi.org/10.1186/1471-2148-12-96.
[4] S.C. Willis, One species or four? Yes! Or, arbitrary assignment of lineages to species obscures the diversification processes of Neotropical fishes, PLoS One 12 (2) (2017), e0172349.

CHAPTER

7

Blue peacock bass, *Cichla piquiti* (Kullander & Ferreira 2006)

Cichla piquiti

Origin of name

Tupi-Guarana adjective "*piquiti*" meaning "striped"

Synonyms

None; this species was not described until 2006

Common names

Blue peacock bass, Striped peacock bass, Tucunaré azul, Tucunaré piquiti

Geographic distribution

Cichla piquiti is endemic to the Tocantins-Araguaia Basin and has been widely introduced in the Upper Paraná Basin and coastal drainages of Brazil's Atlantic Forest region.

Natural history

Found in shallow habitats along the margins of river channels and floodplain lakes and within seasonally flooded areas. Require clear water and have an affinity for submerged structures that provide prey and shelter from predators. Primarily piscivorous.

https://doi.org/10.1016/B978-0-323-85157-2.00012-3

Flying into Palmas, the capital city of the Brazilian state of Tocantins, we (KW, LKW) had a bird's-eye view of the landscape. This region, known as "cerrado," is dominated by tropical savanna, characterized by grasslands with shrubby vegetation. Palmas is infamous for being one of the hottest cities in Brazil, especially during the dry season that extends from May through September. These oppressive temperatures were confirmed as we stepped onto the airport tarmac and felt the hot breeze on our faces. We were met and quickly whisked away by our hosts to our quarters at the Universidade Federal do Tocantins where we had the pleasure of meeting our colleague Fernando Pelicice, professor of fisheries ecology. Among other things, Fernando studies *Cichla piquiti*, the blue peacock bass, in Lajeado Reservoir. The lake was formed in 2002 when a hydroelectric dam was constructed on the Tocantins River, a major river that is sometimes included in maps of the Amazon Basin, even though it flows into a channel that forms only a minor portion of the river's huge estuary. Not only is Fernando an expert on the biology of the blue peacock bass, but he also is an angler skilled in the methods for locating and catching the tucunaré azul, as the fish is called locally. The next morning, equipped with appropriate gear and provisions, Fernando served as our guide and boatman for an excursion on the waters of Lajeado. We cruised across the open waters to an inlet and proceeded to cast lures along the

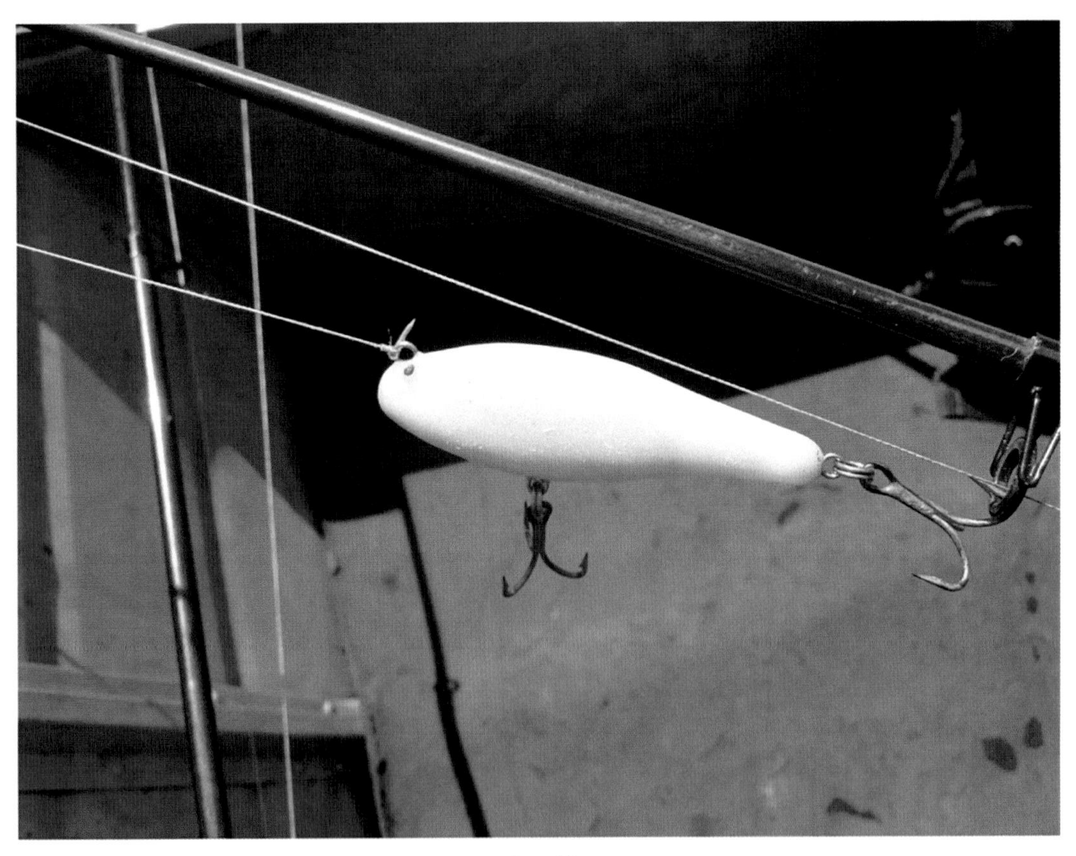

A good lure for catching blue peacock bass. *Photo: K. Winemiller.*

banks near submerged vegetation. Fernando's lure of choice, a white shallow-diving plug with erratic motion, was the winning ticket. In a matter of minutes, his lure enticed a fish to rise from the depths, and then we heard him shout "fish-on!" This lure consistently yielded the "blues" time after time, plus it had the added benefit of rarely snagging when retrieved around sunken logs and branches. This underwater structure provided some of the best habitat for catching both *Cichla piquiti* and the yellow peacock bass, *C. ocellaris* var. *kelberi*.

Professor Fernando Pelicice with a blue peacock bass, *C. piquiti*, taken from Lajeado Reservoir in Brazil. *Photo: K. Winemiller.*

As we reeled in *Cichla piquiti*, we hastily photographed each fish and then passed them to Fernando to measure and record their lengths. He then inserted a numbered "spaghetti tag" to identify each specimen in the event it was recaptured during subsequent surveys. Information from tagged fish can supply a wealth of information for researchers regarding fish growth, survival, and movement. None of the blue tucunaré we landed would be considered trophy size, but we caught many fish that day. Fernando did manage to catch a few fish over 2 kg, and we observed larger fish in the restaurant freezer at the marina. It turns out that Lajeado Reservoir supports not only a fantastic recreational fishery, but also a modest commercial fishery for peacock bass. After just one day of fishing, it was apparent that the reservoir provides a great deal of high-quality habitat for the blue peacock and its smaller cousin, the yellow peacock.

Large male *C. piquiti* from Lajeado Reservoir with a numbered plastic tag used for a mark-recapture study. *Photo: K. Winemiller.*

The blue peacock bass historically occurred only within the Araguaia-Tocantins Basin and was formally recognized by ichthyologists fairly recently with the publication of K&F's monograph [1]. The Araguaia River has a very broad floodplain that contains numerous slow-flowing channels and extensive lakes and wetlands with excellent habitat for peacock bass. In contrast, the Tocantins River has a much steeper bed slope and narrower floodplain, especially in its lower reaches, and consequently contains less habitat suitable for the blue peacock bass, that is until the construction of dams to harness hydroelectricity. While the Tucuruí and Lajeado dams formed barriers to migratory fish and destroyed conditions required by hundreds of river-dwelling fish species, they also created new habitat suitable for the blue peacock bass in the Tocantins River. The proliferation of dams in rivers throughout southern Brazil, the region in the country with the greatest human population and industry, has created reservoirs with conditions favorable for peacock bass. However, the river basins of southern Brazil did not have any native species of peacock bass to take advantage of the new conditions. The solution to this problem, obviously, was to stock reservoirs in the Paraná, São Francisco and other river basins in the south of Brazil with peacock bass native to rivers of the Amazon, and the closest tributaries were the Araguaia and Tocantins. Today, most of the sport fisheries in southern Brazil target the blue peacock bass and yellow peacock bass (*C. ocellaris* var. *kelberi*). Given the proximity of these reservoirs to major urban centers, such as Brasilia, São Paulo, and Rio de Janeiro, it is not surprising that hundreds of photos have been posted on the internet showing proud anglers with their catches of blue peacock bass.

The blue peacock bass, *C. piquiti*, is one of the most distinctively colored peacock bass, but surprisingly was not formally described by ichthyologists until 2006. *Photo: F. Pelicice.*

Identification

Adults and large subadults of the blue peacock bass are easily recognized by having five wide vertical bars (bars 1, 1a, 2, 2a, and 3). These bars extend from the dorsal midline where they are broadest onto the belly region. These bars can vary from faint gray to green-gray to black, but they are always distinct. Other species of peacock bass sometimes express all five (or more) vertical bars, but usually they are not as wide, and in these species the bars are accompanied by other dark pigmentation patterns, such as a horizontal band (*C. intermedia*) or band of blotches on the belly (*C. ocellaris*). An occipital bar on the

Cichla piquiti with blue fins that give this species its common name. *Photo: K. Winemiller.*

A rather drab *C. piquiti* from Lajeado Reservoir that reveals little hint of blue tint in the fins. *Photo: K. Winemiller.*

forehead also is apparent in most *C. piquiti* specimens, but it is much more vertically oriented than the occipital bar sometimes expressed in *C. ocellaris*. Background color of the head and body varies from silver-gray to pale yellow and yellow with a greenish or bluish hue. In some individuals, especially small size classes, there are cloudy white spots scattered all over the head, body, and dorsal fin. The density of these spots varies among individuals, and expression of the trait probably varies according to water quality, other habitat features, and perhaps nutritional or reproductive status. There generally are at least a few black spots or small blotches on the head in the area behind the eye and on the gill cover (operculum), and these often are bordered with white. The caudal ocellus has a white ring, never yellow. The throat and lower portion of the belly and caudal peduncle may be yellow, green, or blue in some adult fish.

Head of *C. piquiti* specimen showing cloudy white spots. *Photo: K. Winemiller.*

The dorsal, caudal, anal, and pelvic fins often are gray with white or light blue spots on the dorsal fin and upper half of the caudal fin. Many individuals have a blue or blue-green tint to the dorsal and caudal fins.

The anal and pelvic fins and the lower half of the caudal fin range in color from yellow to orange to orange-red. In some reproductively active individuals, the entire caudal fin as well as the dorsal, anal, and pelvic fins are blue. This striking blue pigment of the fins is the basis for the common name blue peacock bass. Normally the iris is tan or yellow, but it turns orange in reproductively active individuals. Breeding males develop a nuchal hump that varies considerably in its degree of prominence.

Close-up showing the blue tint of the caudal, anal, and dorsal fin as well as multiple ocelli on the caudal fin and peduncle. *Photo: K. Winemiller.*

Small juvenile *C. piquiti* (less than 8 cm) are very similar in appearance to *C. mirianae* juveniles. They possess a dark lateral band along the entire length of the body, with vertical bars in positions 1, 2, and 3. This lateral stripe may extend onto the postorbital area of the head. In larger juveniles, the body, head, dorsal fin, and upper half of the caudal fin generally are covered with cloudy white spots. As the young fish grow, the characteristic vertical bars become prominent, and the anal fin, pelvic fins, and lower half of the caudal fin may take on a yellow or orange hue.

Close-up of the caudal fin of another *C. piquiti* specimen; this one shows orange in the lower half. *Photo: K. Winemiller.*

Juvenile blue peacock bass in the home aquarium of the authors (KW, LKW).

Juvenile *C. piquiti* from the Upper Paraná River near Porto Rico, Paraná state, Brazil. *Photo: João Birço Latini.*

Distribution and habitat

The blue peacock bass, *Cichla piquiti*, is endemic to the Tocantins-Araguaia Basin. The Tocantins and Araguaia rivers drain the northeastern portion of the Brazilian Shield and merge together before flowing into Bahia de Marajó. The latter is a huge estuarine system that also receives water from the Amazon via channels that pass along the western and southern border of Ilha Marajó to the Atlantic Ocean. The Lower Tocantins (below the merger of the Tocantins and Araguaia) joins the Bahia do Marajó at a location about 65 km southwest of the city of Belém. *Cichla piquiti* is not found in the lowest reaches of the Tocantins or freshwater habitats of the estuary where it apparently is replaced by *C. pinima*. We examined *Cichla* specimens

Geographic distribution of *Cichla piquiti*: blue = native range, purple = areas where the species has been introduced.

archived in the ichthyology collection of Belem's Goeldi Museum that were collected in the estuarine region near Belem, none of which were identified as *C. piquiti*. As mentioned previously, the Tocantins has a steeper slope than the Araguaia. The middle and lower reaches of the Tocantins have areas of swift current and many rocky shoals, whereas the Araguaia has a very shallow slope and is dominated by slow-flowing meandering channels and broad floodplains containing side channels, lakes, and wetlands. About 30 km downstream from the junction of the Tocantins and Araguaia, the Tucuruí Dam was constructed on the Lower Tocantins near the region where a fast-flowing, rocky segment of the river transitions to a broad, slow-flowing sector. The rapids of the Lower Tocantins apparently were a barrier that limited upstream dispersal of other *Cichla* species, except for *C. ocellaris*, which apparently invaded the Tocantins-Araguaia Basin more recently than the ancestor of *C. piquiti*. *Cichla piquiti* is endemic to this basin and distinct from all other *Cichla* species. Conversely, *C. ocellaris* var. *kelberi* is almost indistinguishable from other *C. ocellaris* populations on the basis of morphology and coloration (see Chapter 2), and this stock has certain genetic traits that suggest some degree of interbreeding with *C. ocellaris* var. *monoculus* and perhaps other *C. ocellaris* stocks (see Chapter 11).

In the Araguaia River, *C. piquiti* and *C. ocellaris* var. *kelberi* frequently occur together within the same habitats, but they partition their use of space in a similar manner to peacock bass species that coexist in other regions. The largest blue peacock bass tend to be caught from deeper habitats, including channel areas where the water current is not too swift, compared to areas where *C. ocellaris* var. *kelberi* is common. The latter does not grow as large as *C. piquiti* and generally occupies shallower habitats, especially where there is aquatic vegetation or

Cichla piquiti (top) and *C. ocellaris* var. *kelberi* (bottom) captured from same habitat in Lajeado Reservoir. *Photo: K. Winemiller.*

other kinds of dense submerged structure. The blue peacock bass is found in both rivers and floodplain lakes, but usually is not as abundant in lakes as *C. ocellaris* var. *kelberi*. Juvenile and subadults of both species often occur together in shallow habitats along the margins of river channels and lakes and within seasonally flooded areas. Like all peacock bass, both of these species require clear water and have an affinity for submerged structures that provide prey and also shelter from predators, such as ospreys and otters.

The blue peacock bass has thrived in conditions within the Tucuruí and Lajeado reservoirs, both of which resulted from the impoundment of the Tocantins River for hydropower. Conditions within the reservoirs are highly favorable for proliferation of peacock bass, and both reservoirs now support commercial and sport fisheries for both the blue and yellow peacock bass.

Good habitat for both *C. piquiti* and *C. ocellaris* var. *kelberi* along the shore of Lajeado Reservoir. *Photo: K. Winemiller.*

In the 1990s, hydroelectric companies stocked both species into the Grande, Tietê, Paranaíba and other reservoirs constructed in the Upper Paraná Basin that drains a large portion of southern Brazil [2]. They also were introduced into reservoirs constructed in the São Francisco and other river basins along Brazil's Atlantic coast [3]. These stockings predate K&F's formal description of these species, and early reports describe the blue peacock bass as *C.* sp. "azul" or *C.* cf. *ocellaris*, and the yellow peacock bass as *C. monoculus* or *C.* cf. *monoculus*. These reservoirs

A blue peacock bass captured from Lajeado Reservoir. *Photo: K. Winemiller.*

tend to be fairly nutrient poor with low conductivity and pH from 5 to 7 [4]. Nonetheless, peacock bass quickly established and have become the dominant predatory fishes in constructed lakes and reservoirs throughout east-central and southern Brazil. Genetic analysis indicates there have been multiple introductions of *C. piquiti* into the Upper Paraná Basin [5], and also reveals evidence of hybridization between the blue and yellow peacock bass within reservoirs [5, 6].

Following these interbasin translocations, both peacock bass species spread throughout the Upper Paraná Basin, with dispersal apparently aided by an extensive cascade of reservoirs that provide favorable habitat [2, 7, 8]. Peacock bass have become established as major components of fish communities in floodplain habitats of the Upper Paraná River in Brazil. The rapid spread of peacock bass in the Upper Paraná probably was aided by the reduction in water turbidity (reduced suspended sediment loads) owing to sediment trapping by a series of hydroelectric dams. There are reports that *C. piquiti* also has become established in the Brazilian Pantanal, supposedly introduced after rupture of a stocked pond in the upper Piquiri River. In recent years, anglers have reported catching blue peacock bass from clearwater habitats throughout the Paraguay River system [8]. Sites in the Pantanal where blue peacock bass were captured had temperature ranging from 21 to 32°C, pH from 6 to 6.7, dissolved oxygen concentration from 0.7 to 4.1 mg/L, and transparency from 1.3 to 1.9 m. Thus this species appears to tolerate high temperatures and low dissolved oxygen, but like all peacock bass, requires relatively clear water in order to locate and track prey visually.

Feeding

We are not aware of any diet reports for blue peacock bass from natural habitats of rivers, floodplain lakes, and wetlands within its native range. Fish were the main items found in stomachs of nonnative blue peacock bass from rivers and lakes in the Pantanal [8]. Their diet was dominated by fish in the families Characidae, Cichlidae, and Loricariidae (armored suckermouth catfish) and also included a piranha (*Serrasalmus spilopleura*, Serrasalmidae), headstander (*Leporinus lacustris*, Anostomidae) and a parodontid fish (*Apareiodon affinis*).

The diet and reproductive ecology of the blue peacock bass was investigated in Lajeado Reservoir (Tocantins River) by our host Fernando Pelicice and his colleagues [9]. Over the course of one year, they captured and examined 270 fish and found that all size classes consumed mostly fish, with tetras and cichlids dominating the diet. Silver dollars (*Metynnis* spp.) were the most frequently consumed fish followed by the Araguaia cichlid (*Cichlasoma araguaiense*), small tetras (*Hemigrammus* spp. and *Hyphessobrycon* spp.), and eartheater cichlids (*Satanoperca jurupari*). Also consumed were two piranhas (*Serrasalmus rhombeus*) and three peacock bass, at least one of which was *C. piquiti* (the other two could not be identified). Larger fish tended to eat larger prey.

Small peacock bass are aggressive feeders. This one attacked a bucktail jig. *Photo: K. Winemiller.*

In the Cachoeira Dourada Reservoir in the Upper Paraná Basin, the diet of *C. piquiti* was found to be dominated by fish, including *Satanoperca pappaterra* and other cichlids as well as tetras and other characiforms [4]. Cannibalism also was reported for that population. In addition to fish, aquatic insects and shrimp were reported to be important in the diet of blue peacock bass in other southern Brazilian lakes, and one study reported that cannibalism was predominant [10].

Growth

According to the weight-length relationship calculated for *C. piquiti* from the Cachoeira Dourada Reservoir (Paraná Basin), where surveyed fish ranged from 12 to 27 cm (Paraná Basin), a 27-cm fish should weigh about 5 kg [4]. Both male and female blue peacock bass in Cachoeira Dourada Reservoir attain sexual maturity at about 22 cm and 2 kg.

A study of blue peacock bass from the Volta Grande Reservoir (Paraná Basin) reported a maximum size of 53 cm (in total length, which includes the length of the caudal fin) and 2.8 kg [11], and another study found a maximum length of 54.8 cm [10]. The current IGFA world record for *C. piquiti* is 4.54 kg (10 lbs, 0 oz) taken from the Serra da Mesa Reservoir, ranking the blue peacock bass as 5th largest among the nine species recognized in this book.

Juvenile *C. piquiti* with a dense spotting pattern. *Photo: F. Pelicice.*

A fine specimen of the blue peacock bass from Lajeado Reservoir. *Photo: F. Pelicice.*

Population abundance and structure

An angling survey of *C. piquiti* conducted by Pelicice and colleagues in Lajeado Reservoir produced 119 juveniles and 151 adults of which 132 were male and 109 were female [9]. The fish ranged in size from 12 cm and 55 g to 50 cm and 3.35 kg, and 82% of the fish were smaller than 32 cm. A survey of fish from the Pantanal using both angling and gillnets yielded fish ranging from 15 to 52 cm (total length), with most fish between 32 and 44 cm (weights not reported) [8]. Males were larger (up to 52 cm) than females (up to 45 cm), with most males being from 36 to 42 cm, and most females being from 27 to 36 cm. The sexes had similar abundance in the Pantanal.

A gillnet survey of *C. piquiti* conducted in Cachoeira Dourada Reservoir (Paranaiba River) yielded 102 males and only 39 females [4]. The average standard length of males was 30.5 cm

Cichla piquiti specimens obtained during a survey of Lajeado Reservoir. *Photo: Fernando Pelicice.*

and females averaged 20.5 cm, and weights were not reported. Males and females spanned a similar range of sizes, from 12 to 31 cm. The blue peacock bass was reported to be more abundant in the Pantanal than the yellow peacock bass [6].

In the Itumbiara Reservoir (Paraná Basin), 182 male and 179 female *C. piquiti* were examined during a survey of anglers [12]. The largest male specimen measured 65 cm (total length) and 3.75 kg, and the largest female was 58 cm and 3.25 kg. The average total length of males was 38 cm, and average weight was 0.96 kg. The average length of females was 37 cm and their average weight of was 0.90 kg.

An angling survey of blue peacock bass conducted in the Volta Grande Reservoir produced samples dominated by fish measuring 26–42 cm during the warm, rainy season (October–April) and during the cool, dry season fish measured 30–42 cm [10]. Fishing mortality was estimated to be very low for this population.

Reproduction

There are not any reports of reproduction by native riverine populations of the blue peacock bass, but one study examined reproduction by nonnative fish in the Pantanal [8], and several studies have been conducted in reservoirs. In the Pantanal, *C. piquiti* spawned during the annual low-water period. In Lajeado Reservoir, reproduction appears to take place throughout the year without any synchrony. Dissection and examination of gonads revealed that the average size of testes hardly varied throughout the year, and the average size of ovaries was at a minimum during October and increased gradually to maximum size during March and April. Average body condition (the ratio of body weight to length) did not reveal any seasonal pattern.

In Itumbiara Reservoir, all fish larger than 51 cm and 2.25 kg were sexually mature, and examination of gonads indicated that individual fish spawned multiple times per year in an asynchronous manner [12]. In Cachoeira Dourada Reservoir, half of all males and females were mature at 22.5 cm. Analyses of gonads of *Cichla piquiti* from Cachoeira Dourada Reservoir indicated that at least some fish were reproducing throughout the year, with peak

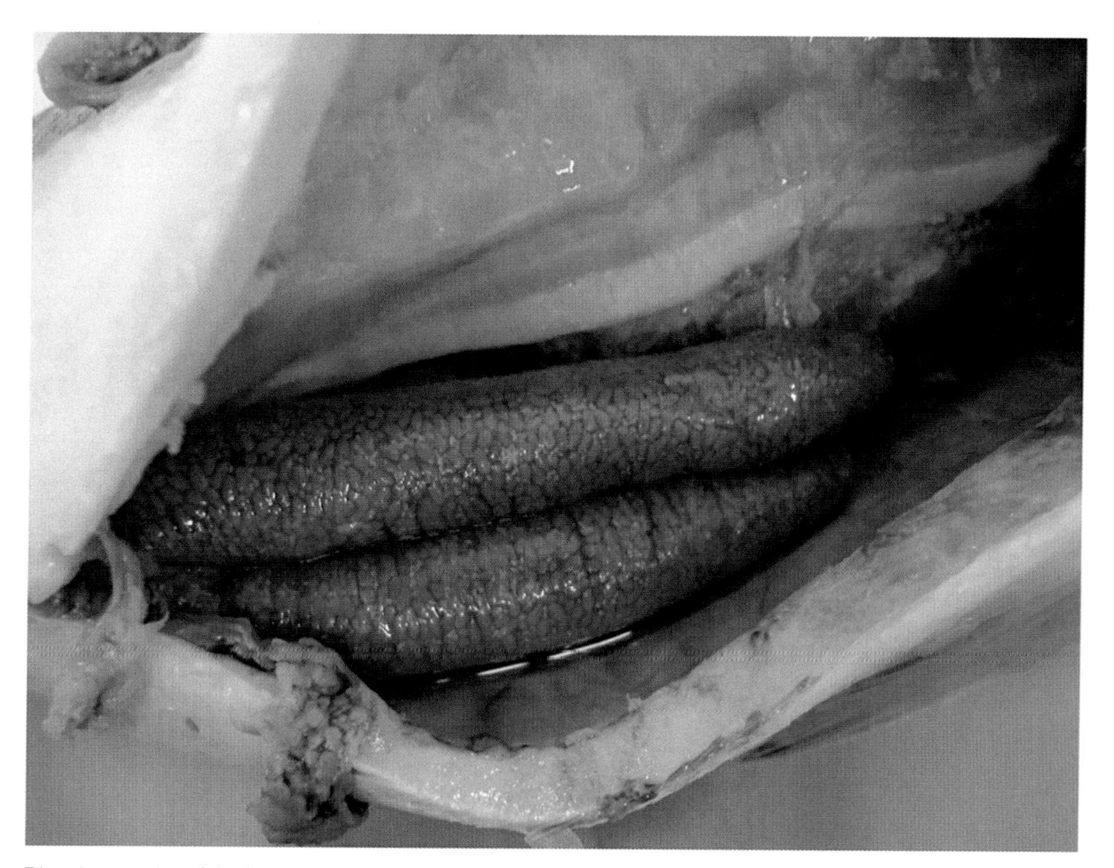

Ripening ovaries of *C. piquiti* from Lajeado Reservoir. *Photo: F. Pelicice.*

spawning during the dry season (April–September) [4]. It seems likely that the blue peacock bass, like other peacock bass species that adapt well to conditions in reservoirs, initiate spawning whenever the individual has obtained a sufficient nutritional state to fuel gonad development and energy expenditures associated with courtship, spawning, nesting, and brood guarding.

Based on a study conducted in the Volta Grande Reservoir, at least some individuals were spawning during all months of the year [11]. Immature fish were collected from January to August with the greatest number observed during February. Mature eggs in ovaries measured 2.3–2.6 mm in diameter. The number of developing and mature eggs in ovaries varied from 3655 to 28,129 eggs per clutch. Larger fish tend to have higher fecundity. The smallest female with ripe ovaries was 34 cm (total length) and weighed 0.7 kg, and the largest was 54 cm and 2.8 kg. Most ovaries contained many more immature than mature eggs, which is indicative of multiple spawning bouts per year.

Males develop a nuchal hump prior to courtship, the size of which varies considerably among individuals. Coloration of both males and females intensifies prior to and during courtship, spawning, and brood care, with the body often taking on a green or green-yellow background coloration, vertical bars becoming darker, the fins turning bright blue, and the iris bright yellow, amber, or orange. The spawning and brooding behavior of the blue peacock bass has been described as being the same as those observed for *C. ocellaris* and other *Cichla* species [11].

Male *C. piquiti* showing blue fins, dark gray bars on a golden background coloration, and a modest nuchal hump. *Photo: F. Pelicice.*

References

[1] S.O. Kullander, E.J.G. Ferreira, A review of the South American cichlid genus *Cichla*, with descriptions of nine new species (Teleostei: Cichlidae), Ichthyol. Explor. Freshw. 17 (2006) 289–398.

[2] A.A. Agostinho, F.M. Pelicice, A.C. Petry, L.C. Gomes, H.F. Julio Junior, Fish diversity in the upper Paraná River basin: habitats, fisheries, management and conservation, Aquat. Ecosyst. Health Manage. 10 (2) (2007) 174–186.

[3] G. Neves dos Santos, A. Filippo, L. Neves dos Santos, F. Gerson Araújo, Digestive tract morphology of the Neotropical piscivorous fish *Cichla kelberi* (Perciformes: Cichlidae) introduced into an oligotrophic Brazilian reservoir, Rev. Biol. Trop. 59 (3) (2011) 1245–1255.

[4] T. Luiz, M. Velludo, A. Peret, J.L. Rodrigues-Filho, A. Peret, Diet, reproduction and population structure of the introduced Amazonian fish *Cichla piquiti* (Perciformes: Cichlidae) in the Cachoeira Dourada reservoir (Paranaiba River, central Brazil), Rev. Biol. Trop. 59 (2011) 727–741.

[5] A.V. Oliveira, A.J. Prioli, S.M.A.P. Prioli, T.S. Bignotto, H.F. Júlio Junior, H. Carrer, C.S. Agostinho, L.M. Prioli, Genetic diversity of invasive and native *Cichla* (Pisces, Perciformes) populations in Brazil with evidence of interspecific hybridization, J. Fish Biol. 69B (2006) 260–270.

[6] L.S. Gasques, T.M.C. Fabrin, D.D. Gonçalves, S.M.A.P. Prioli, A.J. Prioli, Prospecting molecular markers to distinguish *Cichla kelberi*, *C. monoculus* and *C. piquiti*, Acta Sci. Biol. Sci. 37 (2015) 455–462.

[7] A.A. Agostinho, S.M. Thomaz, L.C. Gomes, Conservation of the biodiversity of Brazil's inland waters, Conserv. Biol. 19 (3) (2005) 646–652.

[8] E.K. Resende, D.K.S. Marques, L.K.S.G. Ferreira, A successful case of biological invasion: the fish *Cichla piquiti*, an Amazonian species introduced into the Pantanal, Brazil, Braz. J. Biol. 68 (2008) 799–805.

[9] V.C. de Oliveira Marto, A. Akama, F.M. Pelicice, Feeding and reproductive ecology of *Cichla piquiti* Kullander & Ferreira, 2006 within its native range, Lajeado reservoir, rio Tocantins basin, Neotrop. Ichthyol. 13 (3) (2015) 625–636.

[10] L.M. Gomiero, A.L. Carmassi, G.R. Rondineli, G.A. Villares Junior, F.M.S. Braga, Growth and mortality of *Cichla* spp. (Perciformes, Cichlidae) introduced in Volta Grande Reservoir (Grande River) and in a small artificial lake in Southeastern Brazil, Braz. J. Biol. 70 (4) (2010) 1093–1101.

[11] L. Gomiero, F.M.S. Braga, Feeding of introduced species of *Cichla* (Perciformes, Cichlidae) in Volta Grande reservoir, River Grande (MG/SP), Braz. J. Biol. 64 (2004) 787–795.

[12] A.B.C. Vieira, L.F. Salvador Jr., R.M.C. Melo, G.B. Santos, N. Bazzoli, Reproductive biology of the peacock bass *Cichla piquiti* (Perciformes: Cichlidae), an exotic species in a neotropical reservoir, Neotrop. Ichthyol. 7 (4) (2009) 745–750.

CHAPTER

8

Xingu peacock bass, *Cichla melaniae* (Kullander & Ferreira 2006)

Cichla melaniae

Origin of name

"*melaniae*" in honor of ichthyologist Melanie Stiassny

Synonyms

None; in the decades prior to this species' description, specimens were usually considered to be *Cichla ocellaris*.

Common names

Xingu peacock bass, tucunaré do Xingu, Melanie's peacock bass

Geographic distribution

Cichla melaniae is restricted (endemic) to the lower and middle reaches of the Xingu River and its tributaries, such as the Iriri River.

Natural history

Adapted for living in habitats with clear, flowing water and rocky substrate, residing close to the bottom near submerged logs or rocks. Primarily piscivorous.

Peacock Bass
https://doi.org/10.1016/B978-0-323-85157-2.00007-X

I (KW) was in total disbelief as I gazed upon the Belo Monte Hydroelectric complex. The immense scale of the dam would be hard for anyone to process. How could humans possess the ability to transform a mighty river and the surrounding tropical forests into a world of concrete, mounds of rocks and dirt, and a giant pool of standing water? As an ecologist, it was difficult to accept the reality of imminent destruction of huge spans of rainforest and displacement of 20,000 plus people, many of whom were from indigenous communities. It was August 2015, and I (KW) was participating in a research expedition to document the biodiversity and ecology of the Lower Xingu River, a major clearwater tributary of the Lower Amazon. The Xingu drains the Brazilian Shield, a land formation covering much of the eastern Amazon Basin and a crucible of biodiversity. Our research team consisted of US biologists funded by the National Science Foundation and Brazilian biologists, most of whom, rather ironically, were funded by the energy company developing the hydropower project.

The Xingu River, a major tributary of the lower Amazon, has clear water and flows rapidly through a maze of rocky channels before it flows from the Brazilian Shield into the Amazon lowlands. *Cichla melaniae* and dozens of other fish species occur nowhere else on earth. *Photo: K. Winemiller.*

A Xingu peacock bass from the rocky rapids of the Xingu River, an area now submerged under the reservoir created by the Pimental Dam. *Photo: K. Winemiller.*

Our goal was to document the ecology of as many of the river's aquatic species as possible in the short interval of time before the river was forever transformed by the impending impoundment. We knew from limited surveys conducted previously [1] that the Lower Xingu, and the Volta Grande (Big Bend) stretch in particular, harbor dozens of fish species that occur nowhere else on Earth. The Xingu has very high species endemism (native species found only in a specific area) because of the region's particular geological history and present-day features. The Brazilian Shield has surface formations that date back to the Precambrian (more than 600 million years ago), and the rivers draining this ancient landscape have carved networks of criss-crossing channels into hard igneous and metamorphic rock. Today, the rivers in the upper reaches of the Xingu Basin drain relatively flat terrain and have broad floodplains with rich alluvial soils. As the river begins to descend from the Brazilian Shield before finally emptying its waters into the Amazon, channel slopes and water velocities increase, and there are series of cascades and rapids. Rivers with a steep slope are prime locations for hydropower, because dams can store the kinetic energy of free-flowing water as potential energy in the form of reservoirs that ultimately can convert this stored energy back into kinetic energy driving turbines to produce electricity. In the case of the Xingu River and other Amazon tributaries draining the Brazilian Shield, the natural conditions that are so attractive to hydropower developers are also those required for survival of "rheophilic" fish (fish adapted to live in swiftly flowing water), including many endemic species that are now threatened with extinction.

Belo Monte Dam under construction on the bank of the Xingu River in August 2016. The river can be seen in the background. *Photo: K. Winemiller.*

The zebra pleco, *Hypancistrus zebra*, a species only known to occur in the rapids of the Volta Grande region of the Lower Xingu River. This little fish is highly valued in the ornamental fish trade, but is threatened with extinction due to impacts from hydropower development in the region. *Photo: K. Winemiller.*

The leaders of our expedition were Leandro Sousa, ichthyologist from Brazil's Federal University of Pará, and Mark Sabaj, ichthyologist from the Academy of Natural Sciences of Philadelphia. We had a large team of scientists, students, local boatmen, and local fishermen. The latter make a living diving within the rocky rapids to collect fish for export to the worldwide ornamental fish market. Our ichthyologists and malacologists (scientists who study mollusks) were excited with the prospect of collecting many undescribed species, and these specimens would provide them with research materials for many years to come. I was particularly enthusiastic because I would have my first opportunity to catch *Cichla melaniae*, the Xingu peacock bass, a species that was formally described by K&F [2]. This species is found amid the maze of rocky channels that comprise the Lower Xingu. Attempts with rod and reel rarely failed to catch these feisty predators. I learned to hold the rod tightly, because when one of these fish ambushes a lure, the sudden jolt might dislodge the rod from the grasp of an unprepared angler. Like most peacock bass species, the Xingu peacock has highly variable coloration patterns; however, most of the fish I caught possessed a gray or gray-brown background with barring and spotting patterns that camouflage them within their rocky habitat. It became abundantly clear that the Xingu peacock bass is superbly adapted to the fast-flowing, rocky conditions of the Lower Xingu rapids. Regrettably, much of that habitat has now vanished and been replaced by a huge reservoir with a trickle of water passing downstream.

Identification

According to the species description by K&F, *Cichla melaniae* is distinguished from other peacock bass species "by narrow vertical bars 1-3, and numerous white spots scattered over the flanks, including many ocellated spots." They also note that bars 1a and 2a are commonly present, narrow and indistinct. Actually, in addition to bars 1, 2, and 3, there can be as many as 9 other narrow vertical bars, either faint or quite distinct, located on the side of the body and caudal peduncle. K&F also claim that the species is similar to *C. mirianae*, but light spots are absent from the side of the head; however, in our experience (and in many of the accompanying photographs), this is not the case. The species lacks blotches on the sides of the body, but usually has random scattering of small black spots, with a few black spots widely distributed on the head. In some individuals, these black spots are spread along the body midline to form a pattern reminiscent of a lateral stripe. The caudal ocellus has a white or yellow ring. The background coloration of the body ranges from dull gray to brown, yellow, or green. Like all other peacock bass species, this variation in coloration appears to be associated with age, size, habitat conditions, and the fish's reproductive state. Large adults in breeding condition take on a yellow or yellow-green background color. Males develop a nuchal hump, although apparently not as pronounced as observed in some other peacock bass species.

Cichla melaniae, the Xingu peacock bass, one of only three rheophilic species in the genus. *Photo: K. Winemiller.*

A green halter stripe (diagonal stripe from the corner of the jaws to the lower margin of the cheek) is visible in this beautiful specimen of the Xingu peacock bass. *Photo: M. Sabaj.*

Juveniles smaller than 6 cm have light or silvery background coloration, a faint postorbital stripe, three tiny spots located approximately in the positions of bars 1, 2, and 3, and a small black spot at the base of the caudal fin. Larger juveniles (greater than 10 cm) develop an olive-green or tan background coloration with thin vertical bars in positions 1, 2, and 3, plus thinner and more faint bars in between. There may be cloudy white spots all over the body, and the head lacks any significant blotches or dark spotting. The caudal ocellus begins to take form at about 8–10 cm, but the white ring is usually not complete. The dorsal and caudal fins have translucent white spots, and the pelvic fins may be orange.

Juvenile *Cichla melaniae* measuring approximately 3 cm. *Photo: H. López-Fernández.*

Juvenile *C. melaniae* measuring approximately 8 cm. *Photo: K. Winemiller.*

Distribution and habitat

The Xingu peacock bass, *Cichla melaniae*, is restricted (endemic) to the lower and middle reaches of the Xingu River and its tributaries, such as the Iriri River. This cichlid is most commonly captured from rocky shoals within the network of braided channels that form the lower and middle reaches of the Xingu and Iri rivers, a region where the slope of the landscape increases as the river flows north to the edge of the Brazilian Shield (Guaporé Shield) [1, 3, 4]. Like the Guiana Shield to the north, this formation is dominated by igneous rocks that have been weathered over eons, thus explaining the paucity of suspended sediments carried by rivers of this region. The Xingu peacock bass is not found in the lowermost reaches of the Xingu River where the channel slope is flat and water flow becomes sluggish, nor is it known to occur in the Amazon River into which the Xingu flows. In the lowest segment of the Xingu River, *C. melaniae* is replaced by *C. pinima* and *C. ocellaris* var. *monoculus*, two species that are distributed throughout the lower Amazon River and its tributaries [3, 4]. Therefore it appears that *C. melaniae* is adapted for living in habitats with clear flowing water and rocky substrate, whereas *C. pinima* may occur in more turbid or tea-stained water of slow-flowing river channels or quiet backwaters and floodplain lakes. Measurements of water quality in the Volta Grande rapids region of the lower Xingu indicated slightly alkaline conditions and very high dissolved oxygen content and transparency [3]. The Xingu peacock bass is rare or absent in rivers of the Upper Xingu Basin where the landscape is flat; water flows slowly through channels; aquatic habitats have substrates dominated by silt, clay, sand, and other fine particles; and rocky shoals are rare. In aquatic habitats of upper reaches of the Xingu, *C. melaniae* is largely replaced by *C. mirianae*, the fire peacock bass.

Geographic distribution of *Cichla melaniae* in the Xingu Basin.

Xingu River rapids. *Photo: K. Winemiller.*

Rocky island in a channel of the Xingu River. *Photo: K. Winemiller.*

Excellent habitat for Xingu peacock bass. *Photo: K. Winemiller.*

All size classes of *C. melaniae* are frequently caught near river channel margins where cover is provided by rocks, logs, tree roots, or submerged shrubs. Small juveniles sometimes occur near sandbanks, but larger fish are rarely found over sandy substrate. This species oftentimes is captured by retrieving a lure in swift current that passes near a pocket of still or slow-flowing water that is sheltered from swift current by a large rock or other barrier. These pools hold fish that use a "sit-and-wait" foraging tactic. Thus fish conserve energy by not having to swim in fast water, except for a brief dash to intercept prey swimming along with the current. Because diving plugs with treble hooks tend to snag on rocks, especially when retrieved in fast water, weedless spoons often are an effective lure for anglers targeting these fish. As might be expected of any game fish that inhabits rocky rivers with rapid flows, the Xingu peacock bass strikes lures with great force and has strong endurance for a peacock bass. Kilo for kilo, this is one of the most powerful peacock bass species.

A long stretch of sandbank in the Xingu River. These areas are not good habitat for *C. melaniae*. *Photo: K. Winemiller.*

Feeding

Given that *C. melaniae* was described fairly recently, there has been very little research conducted on the ecology of the species [4, 5]. Because *C. melaniae* has ecological requirements similar to *C. intermedia*, the royal peacock bass, we presume that it forages in the same way and feeds on similar kinds of fish. Both of these species are rarely caught on surface lures, because they normally occupy positions close to the bottom near rocks and submerged logs, making it difficult to present them with a lure. In our experience, best results are obtained for *C. melaniae* when the lure is retrieved in the direction of the water flow (downstream) adjacent to submerged rocks and logs. Weedless spoons and lead jigs work best, because they sink lower in the water column with a slow retrieve and also snag less frequently than lures with treble hooks.

A large pike characin, *Boulengerella cuvieri*, caught from rapids in Lower Xingu River. The fish attacked a shallow diving lure intended for Xingu peacock bass. *Photo: K. Winemiller.*

Growth

Information on the growth or size of maturation has not yet been reported for this species. The Xingu peacock bass probably matures at approximately 30cm at 1–2years of age, which is the size and age reported for most of the smaller peacock bass species. The all-tackle angling record for this species is reported by International Game Fish Association (IGFA) as 3.84kg (8lbs, 7oz), but fish of this size seem to be extremely rare, with most anglers landing fish weighing between 1 and 2kg.

A subadult specimen of *C. melaniae*. *Photo: H. López-Fernández.*

One of several Xingu peacock bass captured from rapids during scientific surveys of the Xingu River. *Photo: K. Winemiller.*

Population abundance and structure

There is one report of seasonal variation in population density in the Volta Grande region of the Lower Xingu River during the years just prior to construction of the Belo Monte Hydroelectric Complex [3]. Fish were surveyed during four phases of the annual flood cycle using experimental gillnets, and density was reported as the number of fish captured per square kilometer of gillnet per hour. Xingu peacock bass captures were greatest during the rising water period when density was reported as 2.3 fish/km^2/hour, followed by the receding water period with 1.6 fish/km^2/hour. Xingu peacock bass were not captured during the period of peak flood stage, and surprisingly, only 0.1 fish/km^2/hour were captured during the period with lowest discharge. Assuming fish should have been most concentrated during the dry season when the water volume is lowest, it seems likely that they were occupying structurally complex habitats within rocky shoals that reduced their vulnerability to capture in gillnets. By comparison, *Geophagus* species (eartheater cichlids) occur over open expanses of sand near shallow banks, and they were captured frequently during the periods of highest and lowest water levels. In these habitats, fish can be captured effectively in gillnets throughout the year. Among the 12 cichlid species captured during their survey, *C. melaniae* was third most abundant after *Geophagus altifrons* and *G. argyrostictus*.

Living in fast-flowing waters, the Xingu peacock bass has power that impresses even the most experienced peacock bass angler. *Photo: K. Winemiller.*

An impressive Xingu peacock bass caught from the Iri River. *Photo: H. López-Fernández.*

Reproduction

We are not aware of any published reports about the reproductive biology of *C. melaniae*, but spawning and nesting behavior are assumed to be similar to that exhibited by other *Cichla* species. Given this species' habitat affinities, it seems likely that it nests amid rocks in areas that are sheltered from strong currents, perhaps in the same manner as we have observed for *C. intermedia* in Venezuela (see Chapter 4). Using a seine net, KW captured juveniles near a sandbank where water flowed slowly over series of shallow ridges and runnels (troughs). Other species captured in these seine samples included the cichlids *Retroclus xinguensis* and *Satanoperca* sp., the benthivorous procholidontid *Semaprochilodus brama*, catfish *Corydoras* sp., and various small tetras (Characidae), the latter probably being a principal prey for small peacock bass.

Male *C. melaniae* from the Iriri River. Note the deep body and prominent nuchal hump of this healthy specimen. *Photo: H. López-Fernández.*

Impacts from hydropower

A major problem confronting the Xingu peacock bass is the destruction of a very significant portion of its natural habitat. Moreover, additional habitat has been severely degraded by dams constructed for the Belo Monte Hydroelectric Complex in 2016 [6]. The Pimental Dam was built just upstream of the Volta Grande on Xingu River near the city of Altamira and flooded an area estimated at 333 km^2. The Pimental reservoir delivers water through a series of canals to a second reservoir that was created on dry land bordering the lower reaches of the Volta Grande. This second reservoir, named Dos Canais, was created by constructing the Bela Vista Dam and a series of dikes. The powerhouse turbines convert up to 11,000 MW from the potential energy stored by these reservoirs. Water is diverted from the Xingu River at a location above the Volta Grande stretch and then returned to the river below the Volta Grande at Belo Monte. Thus the series of rapids that contain some of the best habitat for the Xingu peacock bass consequently have greatly reduced flow, especially during the dry season when most of the available water is diverted for power generation. The dewatered Volta Grande also is being targeted for mining of gold and other valuable minerals. In addition, the Pimental Dam flooded a very long stretch of the Xingu River that contained much excellent habitat for the Xingu peacock bass as well as many other fish species that depend on fast-flowing water and rocky substrates. Like the Xingu peacock bass, many of these "rheophilic" species are endemic to the lower Xingu River. Some peacock bass species thrive in tropical reservoirs; however, *C. melaniae* is not one of these. In fact, there are no records of the royal peacock bass, *C. intermedia*, another species essentially restricted to channel habitats with flowing water and rocky shoals, inhabiting reservoirs or constructed ponds. Time will tell, but it appears unlikely that the Xingu peacock bass will survive in the reservoirs that were created within the heart of this species' native range.

Pimental Dam under construction in August 2016. *Photo: K. Winemiller.*

Dewatered portion of the Xingu River channel below the Pimental Dam during its construction. *Photo: K. Winemiller.*

Habitat for *C. melaniae* and many other endemic fish species of the Xingu River. This area now is submerged by a reservoir. *Photo: K. Winemiller.*

References

[1] J. Zuanon, História natural da ictiofauna de corredeiras do rio Xingu, na região de Altamira, Pará, Instituto de Biologia, Universidade Estuadual de Campinas, Campinas, SP, Brazil, 1999. 198 pp.
[2] S.O. Kullander, E.J.G. Ferreira, A review of the South American cichlid genus *Cichla*, with descriptions of nine new species (Teleostei: Cichlidae), Ichthyol. Explor. Freshw. 17 (2006) 289–398.
[3] T.A. Barbosa, N.L. Benone, T.O. Begot, A. Goncalves, L. Sousa, T. Giarrizzo, L. Juen, L. Montag, Effect of waterfalls and the flood pulse on the structure of fish assemblages of the middle Xingu River in the eastern Amazon basin, Braz. J. Biol. 75 (2015) 78–94.
[4] D.B. Fitzgerald, M.H. Sabaj-Perez, L.M. Sousa, A.P. Gonçalves, L.R. Py-Daniel, N.K. Lujan, J. Zuanon, K.O. Winemiller, J.G. Lundberg, Diversity and community composition of rapids-dwelling fishes of the Xingu River: implications for conservation amid large-scale hydroelectric development, Biol. Conserv. 222 (2018) 104–112.
[5] M.A. Zuluaga Gómez, T. Giarrizzo, D.B. Fitzgerald, K.O. Winemiller, Morphological and trophic diversity of fish assemblages in rapids of the Xingu River, a major Amazon tributary and region of endemism, Environ. Biol. Fishes 99 (8) (2016) 647–658.
[6] M. Sabaj Pérez, Where the Xingu bends and will soon break, Am. Sci. 103 (2015) 395–403.

CHAPTER

9

Fire peacock bass, *Cichla mirianae* (Kullander & Ferreira 2006)

Cichla mirianae

Origin of name

Named after Miriam Leal-Carvalho, who helped collect the specimens used for the species description.

Synonyms

None; in the decades prior to this species' description, specimens from the species range were usually considered to be *Cichla ocellaris*.

Common names

Fire peacock bass, tucunaré fogo, tucunaré vermelho

Geographic distribution

Cichla mirianae is found in river channels and floodplain lakes within the middle and upper portions of the Tapajos and Xingu basins. The species seems to be common in relatively low-gradient rivers and creeks draining forested landscapes of the Upper Tapajos, including the Teles Pires, Juruena, São Benedito, and Arinos rivers. In the Upper Xingu, the species is found in the Batovi, Culuene, and Suiá Missu rivers.

https://doi.org/10.1016/B978-0-323-85157-2.00001-9

Natural history

Occurs in a broad range of habitats, including areas within river channels where the current is not too swift, backwaters, floodplain lakes, wetlands, and low-gradient creeks near submerged structure, especially logs and branches and submerged tree roots, shrubs, grass, and other herbaceous vegetation. Primarily piscivorous.

Nestled in the rainforest along the São Benedito River is the Posada Thaimaçu, a nature lover's haven that teems with monkeys, tapirs, caiman, and a variety of Amazonian bird species. The São Benedito is also an angler's paradise, especially if you're interested in catching the tucunaré fogo, or fire peacock bass, *Cichla mirianae*. I (LKW) along with KW and our son Brent boarded a single engine airplane in Alta Floresta near the northern border of

The São Benedito River is a tributary of the Juruena River (upper Tapajos Basin) that drains pristine tropical rainforest. *Photo: K. Winemiller.*

Brazil's Mato Grosso state. After a short flight over ranchland, recently cleared forest, and ultimately unbroken tracts of pristine rainforest, we touched down on an unpaved airstrip at Thaimaçu. After settling into our cabins, viewing Thaimaçu Falls, and meeting "Fofo" the tapir who visits the camp each night, we prepared our gear for some early morning fishing. At dawn, I set out upstream with Brent and our guide in one boat, while Kirk went with his guide in another boat. Squawking from the dense vegetation along the river banks were dozens of hoatzins, chicken-sized birds reminiscent of the prehistoric *Archaeopteryx*, and lurking just below the water surface with eyes emergent were several large black caimans. We anchored in a lagoon off the main channel and began casting toward submerged brush along a cut bank. Placing a lure among shrubbery without snagging requires a bit of practice, and our guide cheerfully assisted with extraction of a few snagged lures until we became more accurate.

The São Benedito River is a paradise for both naturalists and anglers. The region's rainforest ecosystem is slowly being encroached upon by land development for cattle ranching and crops. *Photo: K. Winemiller.*

Fire peacock bass, *Cichla mirianae*, in the clear water of the São Benedito River. *Photo: K. Winemiller.*

The hoatzin (*Opisthocomus hoazin*), or stinkbird, is found along rivers throughout tropical South America. The chicks of this strange bird have claws on their wings used for climbing. *Photo: K. Winemiller.*

This tapir (*Tapirus terrestris*) was a nightly visitor to Posada Thaimaçu. Note the scars on the cheek and hind leg from a jaguar encounter. *Photo: K. Winemiller.*

Fire peacock bass are plentiful in the São Benedito, and after only a few minutes, Brent and I had landed several fairly small ones along with a couple of pike-cichlids (*Crenicichla* sp.), toothy wolffish (*Hoplias malabaricus*), and of course, the ubiquitous piranhas. After testing our luck at several lagoons, our guide took us to a deep area of the river channel where there was a slow eddy near the bank. We had tried several lures, mostly shallow-diving or surface plugs, either white or yellow to offer some contrast against the dark water. One of my favorite plugs is a white shallow-diver with a red head (a friend aptly named it "the red-headed stranger"). I have several versions of this lure, and in tropical rivers it produces consistently. With the sun rising higher and air temperature along with it, my patience was starting to wane. Nevertheless, I kept up my pursuit of a fish worthy of a photo and future fish tales. Cast after cast, I felt the familiar nibbling of piranhas on the plastic lure. Then, in a flash—wham!—my rod bent downward under the forceful heave of a large fish. I didn't want to lose this one, so I thought to myself—"keep the rod tip up, keep the line taut, let him run, play him, wear him down … get him into the boat." Finally, we saw a crimson body arise from the depths, Brent slipped the landing net under the fish, and I had my prize. This fiery red peacock bass weighed 4kg (8.8lbs). And it did not disappoint—this species lives up to its reputation as the most beautiful of all peacock bass species. Amazingly, this species was not formally recognized as distinct until 2006 when K&F published their taxonomic revision of the genus. Given this recent recognition and the remoteness of the rivers where it occurs, there has been virtually no research done on the ecology of this impressive fish.

A large fire peacock bass from the São Benedito River. *Photo: B. Winemiller.*

Identification

According to K&F [1], *Cichla mirianae* is distinguished from all other *Cichla* species by three large ocellated black blotches along the sides of the body in specimens larger than 9 cm, with adults having black irregular blotches forming a stripe connecting the ocellated blotches and continuing onto the caudal peduncle. This coloration pattern differs from that of *C. orinocensis*, which has three ocellated blotches but lacks the lateral stripe, and *C. intermedia* that has the irregular lateral stripe, but lacks three ocellated blotches. We note, however, that fish from the São Benedito River and Jamanxim rivers often show only scattered irregular black blotches or spots along the sides or else lack any evidence of an irregular black stripe, perhaps because that pigmentation pattern is masked by expression of other colors. This may be the case especially for breeding fish that develop a brilliant background coloration of yellow scales with black edges that produces the appearance of dots and dashes arranged in rows, plus the prominent orange-red or red coloration of the belly and lower sides of the body and head from which the name "fire peacock bass" derives. Fish from other regions of the species' geographic distribution do not seem to express as much of the red coloration or yellow and black dots and dashes pattern as fish from the São Benedito and Jamanxim populations. In some breeding individuals, the black coloration of the three large ocellated blotches may be masked by yellow pigment. The eye is orange in nonbreeding fish and brilliant crimson in breeding fish, and most individuals have bright orange coloration on top of the eyeball that is visible when the fish gazes downward.

In nonbreeding fish, the head and body sometimes are covered with white or cream spots, in the same manner observed in all peacock bass species. This pattern is common in nonbreeding adults in the São Benedito population. In other populations, nonbreeding adults tend to have a uniform yellow or yellow-green head and body along with the three large blotches and small black spots scattered all over the sides. Normally there are a few black spots on the head in the post orbital area, and these may be ringed by cream or yellow pigment. In some specimens, a faint appearance of vertical bars 1a and 2a can be seen. The caudal ocellus usually has a yellow ring. The dorsal fin and upper half of the caudal fin are gray, or sometimes bluish-green, and

Closeup of the head of a fire peacock bass. The orange color on the top of the eyeball is apparent in this photo. *Photo: K. Winemiller.*

Cichla mirianae specimens showing nonbreeding coloration of adults from the Teles Pires (top), Upper Xingu (middle), and São Benedito rivers (bottom two photos). *Photos: top—S. Willis, middle—J. Birindelli, bottom two—K. Winemiller.*

always with rows of light spots. The pelvic fins, anal fin, and lower half of the caudal fin are red, even in nonbreeding adults. Sometimes the anal fin is yellow in breeding adults, or it may have a green hue in nonbreeding fish. These colors may be observed, but to a lesser extent, in the pelvic fins. There often is an ocellated irregular black blotch on the body below the rayed dorsal fin, a pattern and position similar to that frequently found in *C. ocellaris* and *C. pinima*. The distinguishing morphological traits given by K&F, including lateral line features, overlap with those of several other *Cichla* species.

Small juveniles have a dark lateral band along the length of the head and body, with distinct vertical bars in positions 1, 2, and 3. This lateral stripe may extend onto the postorbital area of the head. The bodies and heads of juveniles are often covered with cloudy white or cream spots. Superficially, they appear similar to juveniles of *C. piquiti* and *C. melaniae*.

Juvenile *C. mirianae* specimens collected from the Juruena River in Mato Grosso state, Brazil. These preserved specimens are archived in the Museu de Zoología, Universidade Estadual de Londrina, Brazil. *Photo: K. Winemiller.*

Distribution and habitat

The fire peacock bass is found in river channels and floodplain lakes within the middle and upper portions of the Tapajos and Xingu basins. The species seems to be common in slow-flowing segments of rivers and creeks draining forested landscapes of the Upper Tapajos, including the Teles Pires, Juruena, São Benedito, and Arinos rivers. In the Upper Xingu, the species is found in the Batovi, Culuene, and Suiá Missu rivers. The population in the São Benedito seems to attain the largest sizes (up to 5 kg), with reproductively active fish displaying more brilliant yellow and red coloration when compared to fish from other regions. This perception may be due to the lack of information about fish in breeding condition coming from other regions within the species range. It is also conceivable that the São Benedito population represents a genetically distinct lineage. The genetic data currently available for this species are from fish captured from the Teles Pires River in the upper Tapajos and the Suiá Missu River in the upper Xingu.

Geographic distribution of *Cichla mirianae* in the upper Tapajos Basin.

Excellent habitat for fire peacock bass in the São Benedito River. *Photo: K. Winemiller.*

Sightings of capybara (*Hydrochoerus hydrochaeris*) are not uncommon during fishing excursions for peacock bass in South America. *Photo: K. Winemiller.*

Black caiman (*Melanosuchus niger*) are a common sight in the São Benedito River. *Photo: K. Winemiller.*

The fire peacock bass occurs in a broad range of habitats, including areas within river channels where the current is not too swift, backwaters, floodplain lakes, wetlands, and slow-flowing creeks. Interestingly, the fire peacock bass does not appear to coexist in habitats with other species of *Cichla*, even though its geographic distribution overlaps with *C. ocellaris* var. *monoculus*, *C. pinima*, and *C. melaniae*. This could be because of natural barriers that prevent dispersal, distinct habitat requirements of the species, or perhaps even competition with other *Cichla* species. In our experience, the fire peacock bass is almost always captured near submerged structure, especially logs and branches and submerged tree roots, shrubs, grass, and other herbaceous vegetation. This fish's frequent proximity to woody cover and its impressive strength and endurance on the end of a fishing line, even when compared to other peacock bass species, make landing this fish a challenge. Within seconds of being hooked, these fish generally swim straight to the nearest brush pile or log where the line may become entangled, a tactic also adopted by other peacock bass species. The fire peacock bass appears to be somewhat of a habitat generalist, exploiting shallow edge habitats in a manner similar to *C. ocellaris* and *C. orinocensis*, as well as deeper channel and lake habitats where adults of *C. temensis*, *C. pinima*, and *C. piquiti* often are captured. The fire peacock bass does not appear to inhabit channel areas with swift current and rocks, the kinds of habitats generally preferred by *C. intermedia* in the Orinoco Basin, *C. melaniae* in the Xingu River, and *C. cataractae* in the Essequibo River.

Cichla mirianae taken from the same location in the Teles Pires River, a tributary of the Upper Tapajos. *Photo: S. Willis.*

Feeding

At present there are no published reports on the feeding behavior or diet of *C. mirianae*. The species is presumed to feed nearly exclusively on fish and attacks a variety of artificial lures that mimic fish, including bucktail jigs, diving plugs, surface plugs, and spoons.

These small tetras (*Hyphessobrycon* sp.) are beautiful aquarium fish but also prey for peacock bass. *Photo: K. Winemiller.*

The fire peacock bass will attack virtually any type of lure at the surface or within the water column. *Photo: K. Winemiller.*

The fire peacock bass feeds as aggressively and is as powerful as any of the various peacock bass species. *Photo: K. Winemiller.*

Peacock bass are not the only large predatory fish inhabiting rivers of the upper Tapajos Basin. The toothy peche cachorro (*Hydrolycus scomberoides*) is common in deep, fast-flowing water. *Photo: B. Winemiller.*

Pike cichlids (*Crenicichla* spp.) prey on smaller fish than do peacock bass but are often caught on the same lures used for peacock bass. *Photo: K. Winemiller.*

Growth

No studies have yet been conducted on the growth of the fire peacock bass.

Population abundance and structure

No studies have been done on the abundance, size, or age structure of populations of the fire peacock bass.

The São Benedito has a thriving population of fire peacock bass, the only *Cichla* species found in this river. *Photo: K. Winemiller.*

Rapids on the São Benedito river hold pacus (*Tometes* sp.), a relative of the piranha that can give quite an exciting experience for anglers using light gear in the fast water. *Photo: K. Winemiller.*

Reproduction

To date, nothing has been documented regarding the reproductive ecology of the fire peacock bass. Males in breeding condition develop a nuchal hump, although it does not appear to be as exaggerated as in species such as *C. ocellaris*, *C. intermedia*, and *C. melaniae*. Both males and females assume the brilliant yellow and black spotted background coloration of the upper portions of the head and body, with fiery orange-red pigment splashed across the head from the lips to the throat (including branchiostegal membranes) to the eye and postorbital

A healthy male *C. mirianae* in breeding condition, showing bright yellow, black, and fiery red coloration and nuchal hump. *Photo: K. Winemiller.*

Fire peacock bass from the Jamanxim River—male showing nuchal hump. Note that the anal fin has been extensively nipped by piranhas. *Photo: J. Birindelli.*

area. This orange-red pigmentation often covers the entire lower half of the body, the entirety of the pelvic fins and anal fin, and lower half of the caudal fin. The anal fin of some breeding adults is more yellow than orange or red. One presumes that many of the brilliantly colored fish that strike lures are defending nests or broods. Many photos posted on the internet show brilliantly colored fish in a pre- or postspawning reproductive state, and they likely are fish captured from the São Benedito and Jamanxim rivers.

Fire peacock bass from the São Benedito River. This robust fish likely is a female that is approaching reproductive condition. *Photo: K. Winemiller.*

A fire peacock bass nest observed on the sandy bottom of the São Benedito River. *Photo: K. Winemiller.*

A magnificent fire peacock bass lives to fight another day. *Photo: K. Winemiller.*

Reference

[1] S.O. Kullander, E.J.G. Ferreira, A review of the South American cichlid genus *Cichla*, with descriptions of nine new species (Teleostei: Cichlidae), Ichthyol. Explor. Freshw. 17 (2006) 289–398.

CHAPTER

10

Falls lukunani, *Cichla cataractae* (Sabaj, López-Fernández, Willis, Hemraj, Taphorn & Winemiller 2020)

Cichla cataractae

Origin of name

Species name is derived from *cataractae*, Latin for waterfall or rapids.

Synonyms

None; until very recently, ichthyologists confused this species with *Cichla ocellaris*, a species that is superficially similar and with which it coexists within the Essequibo Basin in Guyana.

Common names

Falls lukunani, a name that reflects this species' affinity for water flowing over rocks.

Geographic distribution

The falls lukunani is only known to occur in the Essequibo River, including its major tributaries such as the Cuyuni and Rupununi rivers.

Natural history

Associated with rocky shoals in flowing channels of clearwater and moderate blackwater rivers. Primarily piscivorous.

https://doi.org/10.1016/B978-0-323-85157-2.00004-4

For the past 2 weeks, I (KW) had been participating in an expedition to survey fish in the Rewa River in a remote area in southern Guyana blanketed by rainforest. During the final 2 days of our expedition, I would at last have the opportunity to catch the falls lukunani. The expedition was led by my former graduate student and colleague Dr. Hernán López-Fernández, who currently is a professor and curator of ichthyology at the University of Michigan. Other members of the team included Don Taphorn, the ichthyologist who, in the mid-1980s, first introduced me to South American fishes in the Venezuelan Llanos when I was a graduate student, Devin Bloom, an ichthyologist from Western Michigan University, plus two graduate students and a postdoc working with Hernán and Devin. After viewing photos posted on the internet, I knew there was a unique peacock bass being caught by anglers in Guyana, and I strongly suspected it was an undescribed species that people were confusing with another species of lukunani, *Cichla ocellaris*. The lukunani (butterfly peacock bass, Chapter 2) is a common and much-sought-after sport fish in Guyana, Suriname, and French Guyana.

I was disappointed when our local guide, Ashley Holland, informed me that the falls lukunani did not occur in the Rewa River and could only be found downstream in the Rupununi and Essequibo rivers. He explained that local people were well aware of two kinds of peacock bass in the Essequibo, the pond lukunani (*C. ocellaris*) and the falls lukunani (the form I was seeking). He added that the falls lukunani is always captured on rocky shoals in the river channel, whereas the pond lukunani is taken from a variety of habitats, for example adjacent to channel banks, backwaters, floodplain lakes, creeks, and even constructed ponds and canals. This new information only strengthened my suspicion that the falls lukunani is a separate species.

After we had descended the Rewa to spend two final nights camped along the Rupununi River, we finally would have our opportunity to obtain specimens of the falls lukunani. The first evening we motored to the nearest patch of submerged rocks and began casting lures. In no time at all, one of our party captured a healthy falls lukunani, a fish instantly recognized as distinct from the many pond lukunani that we had caught from the Rewa River during the previous days. As day turned to night, we returned to camp with just a single specimen. The next morning, we returned to the rocky shoals and caught three more falls lukunani in addition to a couple of pond lukunani, black piranhas (*Serrasalmus rhombeus*), and a large arowana (*Osteoglossum bicirrhosum*). With just four specimens of the falls lukunani in hand, we had to break camp and rush downriver to meet our airplane to fly back to Georgetown.

Analysis of DNA from tissue samples from the four specimens performed by our colleague Stuart Willis indicated that the falls lukunani is more closely related to *Cichla temensis* and *C. pinima* than to *C. ocellaris*, the pond lukunani. Ichthyologists had considered these two species as one and the same for more than two centuries. Even the legendary Neotropical ichthyologist Carl Eigenmann, who collected extensively in the Essequibo River, had failed to recognize the distinct morphological, coloration, and ecological differences between the falls lukunani and pond lukunani [1]. Mark Sabaj, an ichthyologist at the Academy of Natural Sciences of Philadelphia, discovered some additional archived specimens, including juveniles, in the fish collection at his museum. A formal taxonomic description of the falls lukunani was published in the journal *Proceedings of the Academy of Natural Sciences of Philadelphia* [2]. The authors are Mark, Hernán, Don, Stuart, myself (KW), and Devya Hemraj, a graduate student from the University of Guyana who assisted with several ichthyological expeditions in Guyana.

One of four falls lukunani (*Cichla cataractae*) caught from rocky shoals of the Rupununi River during an ichthyological survey conducted in 2018. *Photo: K. Winemiller.*

Following results from genetic studies [3, 4], the 15 *Cichla* species identified by K&F [5] condense to just eight species that we now recognize (see Chapter 11). The recent description of *C. cataractae* brings the number of recognized peacock bass species to nine. It certainly is possible that future biological surveys within the vast Amazon and Orinoco basins, along with examination of larger numbers of specimens and additional genetic analyses, may identify a few more undescribed species of peacock bass.

Falls lukunani (*Cichla cataractae*) from the Essequibo River, Guyana. *Photo: A. Holland.*

Identification

Cichla cataractae and *C. ocellaris* are similar morphologically and have some similar coloration patterns that vary depending on environmental conditions and the fish's reproductive state, but they also have some consistent differences in pigmentation. In some respects, the coloration patterns of the falls lukunani share greater resemblance to those of *C. pinima*.

The background coloration of the head and body varies from gray-green to yellow-green or bronze-green with a white belly. Sometimes there is a very faint appearance of up to 12 thin vertical bars extending from the top of the body to the lateral line. The caudal ocellus has a bright yellow or white ring. Some adult specimens have evenly spaced white spots covering the posterior half of the body, with rows of white spots covering the rayed dorsal fin and upper half of the caudal fin, and both of these fins are gray with a blue or blue-green tint. The bottom half of the caudal fin, anal fin, and pelvic fins are yellow-orange or orange.

The most distinctive feature of the falls lukunani is a large, round, ocellated blotch at the midline in the area where vertical bar 1 normally occurs on some peacock bass species. In addition to this characteristic marking, the falls lukunani also possesses a smaller and more irregular-shaped ocellated blotch on the body just below the rayed dorsal fin. Sometimes there is a second smaller blotch adjacent to the spot just below the rayed dorsal fin. An ocellated blotch below the rayed dorsal fin also is commonly observed in *C. ocellaris*, *C. pinima*, and *C. mirianae*. The falls lukunani occasionally has a vertical blotch just below the large blotch in the position of bar 1. In some individuals, there are additional ocellated small blotches that merge together to form a thin irregular vertical bar in the position of vertical bar 2. Some individuals lack blotches forming thin vertical bars, a condition that may vary

Three specimens of the falls lukunani captured from the same rocky shoal in the Rupununi River. Variation in coloration patterns is apparent, but several unique features are consistent in all three specimens. The bottom fish is a male in breeding condition, with a small nuchal hump and bright red iris. The middle fish likely is a gravid female. *Photos: K. Winemiller.*

according to the fish's reproductive state. The falls lukunani never shows evidence of the irregular black blotches in the belly region that are a distinguishing characteristic of *C. ocellaris*. Males develop a nuchal hump, but it does not seem to be as pronounced as those observed in many other *Cichla* species. The lower region of the head, throat, chest, and belly is sometimes yellow, yellow-green, or orange hue, and this coloration may be dependent on water quality and/or reproductive state. The iris of the eye ranges from orange to crimson, with the latter apparently indicative of spawning preparedness, active nesting, and brood guarding. There generally are at least a few black spots in the postorbital region of the head, but more frequently there are black blotches, and these generally are ringed with white, cream, or yellow. Black spots and blotches on the head are another feature that distinguishes the falls lukanani from *C. ocellaris*.

This subadult falls lukunani was caught from the Cuyuni River in Venezuela. *Photo: C. Montaña.*

Juvenile falls lukunani differ markedly from those of *C. ocellaris*. Juvenile falls lukunani (5–10cm) have prominent black blotches on the head that form a broken horizontal stripe in the postorbital region, whereas juvenile *C. ocellaris* lack any dark pigmentation on the head. There also is a faint stripe running from the tip of the snout to the eye in juvenile falls lukunani. Furthermore, these juvenile fish have a large blotch in the position of bar 1; a very small spot or small vertical blotch in the position of bar 2; and a larger, elongate, triangular blotch running horizontal near the midline below the rayed dorsal fin, extending from the position of bar 3 onto the caudal peduncle. Juveniles also have a large blotch in the position of the caudal fin ocellus that runs into the area just anterior to this spot. In juveniles larger than 10cm, these blotches have a white border. In contrast, juvenile *C. ocellaris* have three small spots along the body midline, and a thin horizontal stripe that begins at the third spot and ends at a small spot at the base of the caudal fin. Juveniles smaller than 5–6cm have a silver background coloration with a blue-green sheen, and larger juveniles transition to beige or tan and then develop the background coloration as described for adults. Juvenile falls lukunani may possess up to 10 faint thin vertical bars along the body that extend from the dorsal area almost to the belly in some cases. The dorsal and caudal fins of juveniles are darkly pigmented with large transparent spots.

Juvenile falls lukunani (*Cichla cataractae*). The top photo is a preserved specimen (13.8 cm) from the Essequibo River archived in the Ichthyology Collection of Academy of Natural Sciences of Philadelphia, and the bottom photo is specimen (8.2 cm) collected from rapids in the Cuyuni River in Venezuela and archived in the Ichthyology collection of the Museo de Historia Natural de Guanare in Venezuela. K&F recognized that the Cuyuni specimen was not *C. ocellaris*, and tentatively identified it as *C.* cf. *orinocensis*, a species that does not naturally occur in the Cuyuni-Essequibo Basin. *Top photo: M. Sabaj; bottom photo reproduced from K&F 2006.*

For comparison, this photo shows two juvenile pond lukunani (*C. ocellaris*) captured from the Rewa River where the falls lukunani is absent. Note there are no spots, blotches, or stripes on the head. *Photo: K. Winemiller.*

Falls lukunani (*Cichla cataractae*, top) and pond lukunani (*C. ocellaris*, bottom) caught from the same location. Both are males with nuchal humps. These fish are superficially similar, but the top fish has small black spots on the head and lacks abdominal blotches and orange-red coloration in the throat and belly region as well as the pelvic, anal, and caudal fins. Although both fish have an ocellated blotch below the rayed portion of the dorsal fin (actually two small blotches in the top fish), the top fish also has a round blotch near the body midline in the position of vertical bar 1, whereas the bottom fish has a round blotch located more dorsally on the body in position of vertical bar 1, plus an additional blotch in the position of vertical bar 2. *Photo: K. Winemiller.*

Distribution and habitat

The falls lukunani is only known to occur in the Essequibo River, including its major tributaries such as the Cuyuni and Rupununi rivers. The species has been captured from the Cuyuni River in Venezuela and appears to be common within suitable habitats within the Essequibo Basin in Guyana. The species is apparently restricted to rivers with low to moderate gradients, and it seems to be absent from upper reaches of Essequibo tributaries, such as the Rupununi and Rewa rivers. The falls lukunani is strongly associated with rocky shoals in flowing channels of clearwater and moderate blackwater rivers, and this affinity for water flowing over rocks is the reason for its local name. Adult and subadult falls lukunani are not normally found in floodplain habitats, but it is likely that fish may enter flooded marginal areas during some periods and life stages, for example juveniles seeking food and refuge from predation.

Map showing the distribution of the falls lukunani (*Cichla* cataractae) in Guyana and Venezuela.

Rocky shoals in the Rupununi River provide suitable habitat for the falls lukunani. *Photo: K. Winemiller.*

Falls lukunani (*Cichla cataractae*) captured from the Essequibo River near the ferry crossing on Linden-Lethem road. *Photo: D. Taphorn.*

Feeding

There are not any published reports about feeding behavior or diet of the falls lukunani. Undoubtedly it feeds primarily or exclusively on fish, and anglers catch them using spoons, surface and diving plugs, spinner baits, bucktail jigs, and streamers. Fry are presumed to feed on zooplankton and other small aquatic invertebrates before shifting to a diet of fish.

Local guide Ashley Holland with a beautiful falls lukunani from the Essequibo River. *Photo: A. Holland.*

Growth

Information related to growth is not available for this species that only recently was recognized as being distinct from *C. ocellaris*. Local guide Ashley Holland of Yupukari village told us that the falls lukunani grows larger than *C. ocellaris*, with individuals of the former weighing up to 5 kg, and the latter rarely attaining 4 kg.

A large falls lukunani (*Cichla cataractae*) from the Essequibo River, Guyana. This male has a bright red iris and well-developed nuchal hump. *Photo: A. Holland.*

Population abundance and structure

Reports containing abundance or size and age structure information have not been published for this species, but the falls lukunani (*C. cataractae*) appears to be less abundant than the pond lukunani (*C. ocellaris*) in rivers where the two co-occur. However, when fishing around rocky shoals in appropriate river reaches, the falls lukunani seems to be most common, whereas a few *C. ocellaris* may be caught near the shoreline or from habitat patches well sheltered from water current.

Professor Donald Taphorn with a falls lukunani captured in a gillnet from the Essequibo River. *Photo: D. Taphorn.*

Reproduction

Information is not available regarding falls lukunani reproductive habits. Given its habitat affinity, it is presumed to nest on rocky shoals in the river channel during seasons of low discharge in a manner similar that observed for *C. intermedia* in Venezuela. Male falls lukunani develop a nuchal hump prior to breeding, but these tend to be small in all specimens observed by us in the field or in photos posted on the internet.

These specimens likely represent nonbreeding (top) and breeding (bottom) falls lukunani. *Photo: K. Winemiller.*

Shallow areas in rocky shoals that are sheltered from strong current likely are nesting sites for the falls lukunani. *Photo: K. Winemiller.*

Juvenile *C. cataractae* caught with a seine net near the shore of the Essequibo River. *Photo: Wes Wong.*

References

[1] C.H. Eigenmann, The freshwater fishes of British Guiana, including a study of the ecological grouping of species, and the relation of the fauna of the plateau to that of the lowlands, Mem. Carnegie Mus. 5 (1) (1912) 1–103.

[2] M.H. Sabaj, H. López-Fernández, S.C. Willis, D.D. Hemraj, D.C. Taphorn, K.O. Winemiller, *Cichla cataractae* (Cichliformes: Cichlidae), new species of peacock bass from the Essequibo Basin, Guyana and Venezuela, Proc. Acad. Natl. Sci. Phila. 167 (2020) 69–86.

[3] S.C. Willis, J. Macrander, I.P. Farias, G. Orti, Simultaneous delimitation of species and quantification of interspecific hybridization in Amazonian peacock cichlids (genus *Cichla*) using multi-locus data, BMC Evol. Biol. 12 (2012) 96.

[4] S.C. Willis, I.P. Farias, G. Orti, Multi-locus species tree for the Amazonian peacock basses (Cichlidae: *Cichla*): emergent phylogenetic signal despite limited nuclear variation, Mol. Phylogenet. Evol. 69 (3) (2013) 479–490.

[5] S.O. Kullander, E.J.G. Ferreira, A review of the South American cichlid genus *Cichla*, with descriptions of nine new species (Teleostei: Cichlidae), Ichthyol. Explor. Freshw. 17 (2006) 289–398.

CHAPTER

11

Evolutionary relationships and zoogeography

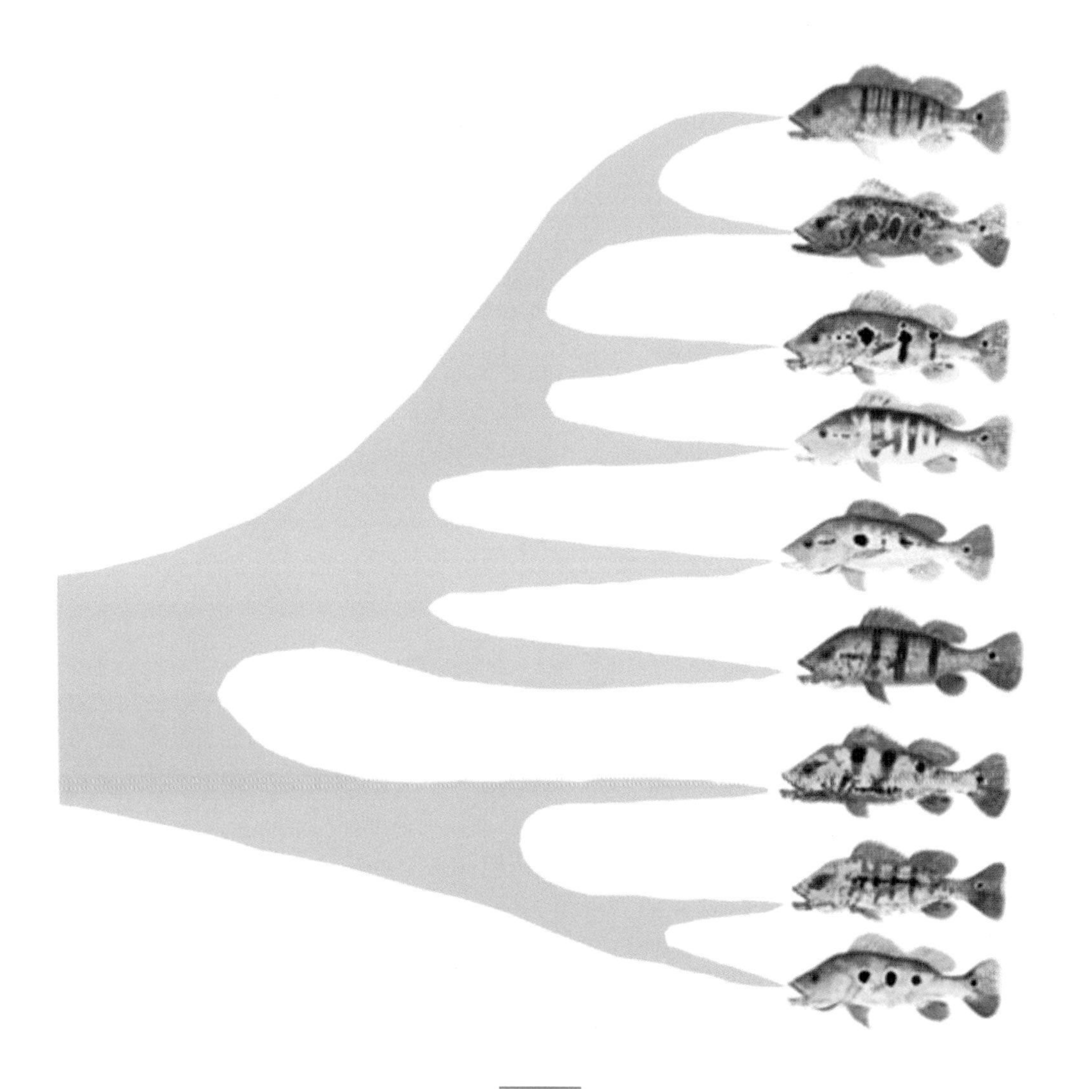

Peacock Bass
https://doi.org/10.1016/B978-0-323-85157-2.00003-2

As discussed in Chapter 1, peacock bass belong to the Cichlidae, a family of tropical freshwater fishes with over 1600 described species and perhaps more than 3000 species if we include the hundreds yet to be described. The English common name "bass" is used for a variety of freshwater and marine fish belonging to various families, such as largemouth and smallmouth bass from freshwaters of North America (Centrarchidae), striped bass and sea bass from the North Atlantic (Moronidae), and Chilean sea bass off the coast of Patagonia (Nototheniidae). Cichlids are extremely diverse in South America (more than 400 described species), Central America (about 160 species), and Africa (well over 1000 species, most of which are endemic to the Great Lakes of the East African Rift Valley), and there are a few species found in Cuba and Hispaniola (*Nandopsis* with 3 species), Madagascar (6 genera, several species), India (2 genera, 3 species), Iran (*Iranocichla* with 2 species), and the Middle East (6 genera, 9 species). Cichlids are placed within the class Actinopterygii (ray-finned bony fishes) and order Cichliformes, a lineage containing only the Cichlidae and Pholidichthyidae (convict blennies from the tropical Pacific Ocean).

A feature that distinguished peacock bass from other Neotropical cichlids is their relatively large mouth. *Photo: C. Venegas.*

Cichla are distinguished from other South American cichlid genera by their large mouth, forward projected lower jaw, notched dorsal fin (spine length increases until the 5th spine then decreases to a short penultimate spine in front of the posterior rayed portion), and scaly fins (except for the pectorals), among other traits [1]. The dimensions of the head, mouth, body, and fins that influence swimming performance and feeding ability are very similar among the various peacock bass species. This suggests either that evolutionary divergence among *Cichla* species is fairly recent without sufficient time for major morphological changes, or that the ecological niche, or role, of peacock bass has been highly conserved over an extended period of evolution. Early anatomical studies estimated that *Cichla* is most closely related to the South American pike cichlids (*Crenicichla* species) and positioned near the root of the Neotropical cichlid evolutionary tree [2–4]. However, more recent studies based on analysis of DNA sequence data have determined that all Neotropical cichlids descended from a common ancestor and that *Cichla* is more closely related to *Retroclus* species rather than pike cichlids [5, 6].

Retroclus xinguensis, a cichlid from the Xingu River that feeds on benthic invertebrates, belongs to the genus of fishes now considered to be most closely related to peacock bass. *Photo: K. Winemiller.*

Crenicichla lenticulata, a pike cichlid, a member of the lineage that was formerly estimated to be most closely related to peacock bass until new genetic studies advanced our understanding of evolutionary relationships. *Photo: K. Winemiller.*

As mentioned in preceding chapters, the taxonomic revision by K&F [7] identified 15 *Cichla* species based on evaluation of anatomical and coloration patterns that sometimes are unique but often overlap between species. More recent analyses of peacock bass lineages based on genetic data [8–10] have identified only eight species among the 15 proposed by K&F. Chapters 2–10 discussed the taxonomy, unique coloration patterns, and geographic distribution of these eight species plus the very recently described falls lukunani (*C. cataractae*) from the Essequibo Basin. Here we discuss evolutionary relationships within the genus *Cichla* and review some recent genetic evidence regarding hybridization.

Evolutionary relationships

Our overview of the evolutionary relationships of peacock bass species and varieties is based on the population genetics and phylogenetic analyses completed by Dr. Stuart Willis, currently the world's foremost expert on the genetics of these fish. As a boy, Stuart was a tropical fish hobbyist who eventually mastered the rearing of discus fish (*Symphysodon* species), cichlids that require specialized care with respect to water quality and dietary needs. Stuart began his college career at Texas A&M University where he collaborated with us on research projects related to tropical fish ecology. He later pursued advanced degrees at the University of Manitoba and University of Nebraska, where he trained in molecular techniques while initiating his research on the diversity and evolutionary relationships of peacock bass. With the help of many colleagues, including ourselves but especially scientists and students in Brazil, Stuart obtained peacock bass from rivers throughout the Orinoco and Amazon basins. His samples included fish from nearly all the localities mentioned in K&F's monograph plus many additional sites. Using tissue samples from these specimens, he extracted and amplified DNA for nucleotide sequencing and statistical analysis to reveal genetic similarities and differences among individuals and populations. The scientific fields of molecular phylogenetics (evolutionary relationships based on analysis of patterns of genetic variation, i.e., differences among DNA nucleotide sequences) and genomics (determination of gene structure, regulation, expression, and interactions based on computational analyses involving large amounts of nucleotide sequence data) are advancing rapidly, and it is beyond the scope of this book to delve into details of Stuart's methods. Here we will summarize his main findings and conclusions about the evolutionary relationships of peacock bass [8–13]. Now Stuart is investigating the genetics of salmon and other marine fishes, while continuing efforts to advance knowledge of peacock bass genetics.

Based on phylogenetic analysis of molecular data, Willis and colleagues identified two major evolutionary units (clades, which are lineages sharing a common ancestor) within the genus *Cichla*. They called the larger unit "Group A," a clade containing *C. temensis*, *C. pinima*, *C. piquiti*, *C. melaniae*, and *C. mirianae*. This clade includes the peacock bass species that attain the largest sizes. "Group B" contained three of the smaller species—*C. ocellaris*, *C. intermedia*, and *C. orinocensis*. A recent analysis performed by Willis and colleagues using the same methodology, but including recently acquired tissue samples from the falls lukunani, *C. cataractae*, produced the same pattern of relationship among those eight species and placed the falls lukunani within Group A.

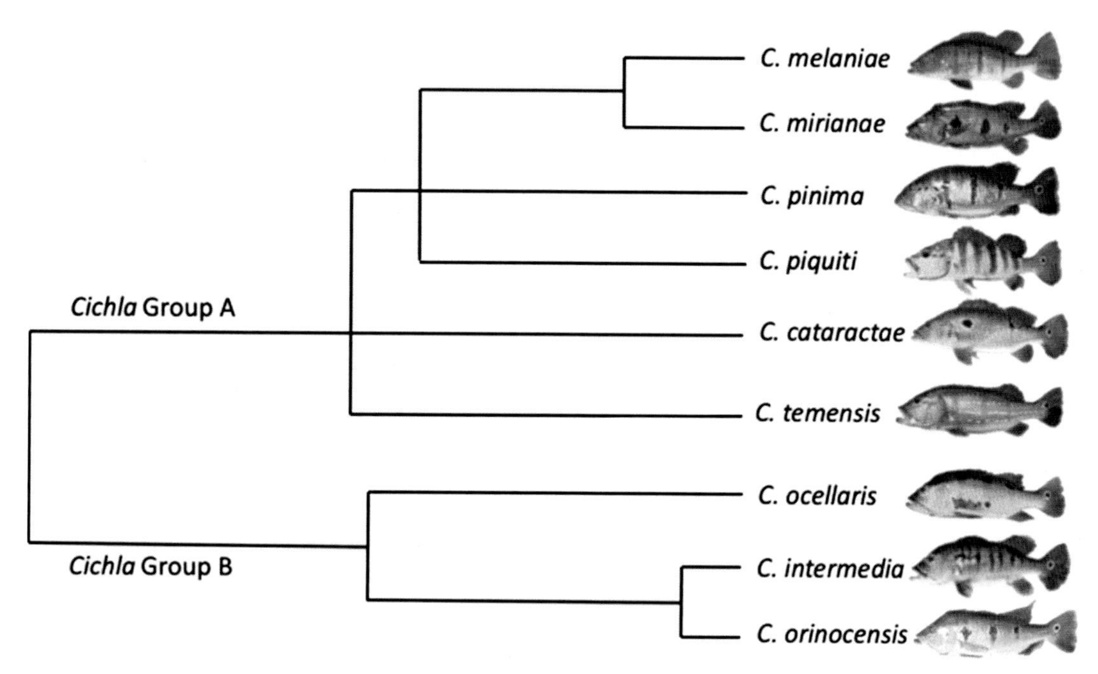

Evolutionary relationships of *Cichla* species based on analysis of DNA sequence data [8, 9].

Within Group A, the sequence of evolutionary divergence is unresolved with regards to *C. temensis*, *C. cataractae*, and the common ancestor of the other four species in this group. *Cichla melaniae* and *C. mirianae* are sister species (i.e., they share a common ancestor), and the succession of divergence is unresolved for the clade containing the lineages for *C. piquiti*, *C. pinima*, and the ancestor of *C. melaniae* and *C. mirianae.* In Group B, *C. intermedia* and *C. orinocensis* share a common ancestor that split from the *C. ocellaris* lineage.

Analysis of genetic variation in *C. temensis* revealed high consistency throughout its geographic distribution indicative of a single species, but also small genetic differences among regional stocks that suggest possible adaptation to local environmental conditions [12]. Analysis of genetic variation in *C. pinima* presents a more complicated scenario [13]. Patterns of genetic divergence among regional stocks suggest the existence of perhaps four "lineages," none of which corresponds to the four species described by K&F and that we treat here as a single species, *C. pinima*. Fish identified by K&F as *C. jariina*, *C. thyrorus*, and *C. vazzoleri* are

Distribution map adapted from a figure in [13] showing essentially nonoverlapping geographic ranges of species that constitute clade A from a molecular phylogeny of *Cichla*.

Cichla melaniae belongs to the Group A lineage along with *C. mirianae*, *C. pinima*, *C. piquiti*, *C. temensis*, and *C. cataractae*.

not genetically distinct from stocks they identified as *C. pinima*. This is an interesting finding, because the *C. jariina* stock is found in the Jari River upstream from major cataracts, and yet genetically, it is almost identical to *C. pinima* from the lowlands of the eastern Amazon. Either the cataracts are not an effective barrier to dispersal and gene flow (interbreeding of stocks), or else the barrier was formed so recently that there was insufficient time for stocks above and below the cataracts to diverge genetically. However, there are genetically divergent *C. pinima* stocks in other regions of the Amazon that Willis describes as a "Western core," "Southern core," "Aripuanã-Machado," and "Jatapu." These populations are not geographically isolated from one another. However, the situation is complex because fish from the easternmost Amazon (region downstream from the Tapajos mouth) are the product of dispersal and interbreeding of fish from both the Southern and Western core stocks. Genetic variation among these various stocks implies reproductive isolation, but the occurrence of shared nucleotide sequences also suggests recent interbreeding among stocks. Whether *C. pinima* should be considered a single species, two species (Western vs. Southern), or four species with a zone of intermixing in the easternmost Amazon is debatable [13]. Since there are not any unique anatomical or coloration characteristics that can consistently separate these stocks from one another, and because the ecology of this species seems to be consistent throughout its range, we consider *C. pinima* to represent a single species with a broad geographic distribution, second in size only to that of *C. ocellaris*. At the same time, we acknowledge that some stocks sometimes have distinct genetic patterns that may reflect some degree of adaptation to local environmental conditions in a manner similar to *C. temensis* and perhaps all peacock bass species and varieties.

Cichla pinima from the Jari River. Based on analysis of morphology and coloration patterns, fish from this region were given the name *C. jariina* by K&F, but subsequent genetic analyses indicate that they are not different from *C. pinima*, a species broadly distributed in lowlands of the eastern Amazon Basin. *Photo: S. Willis.*

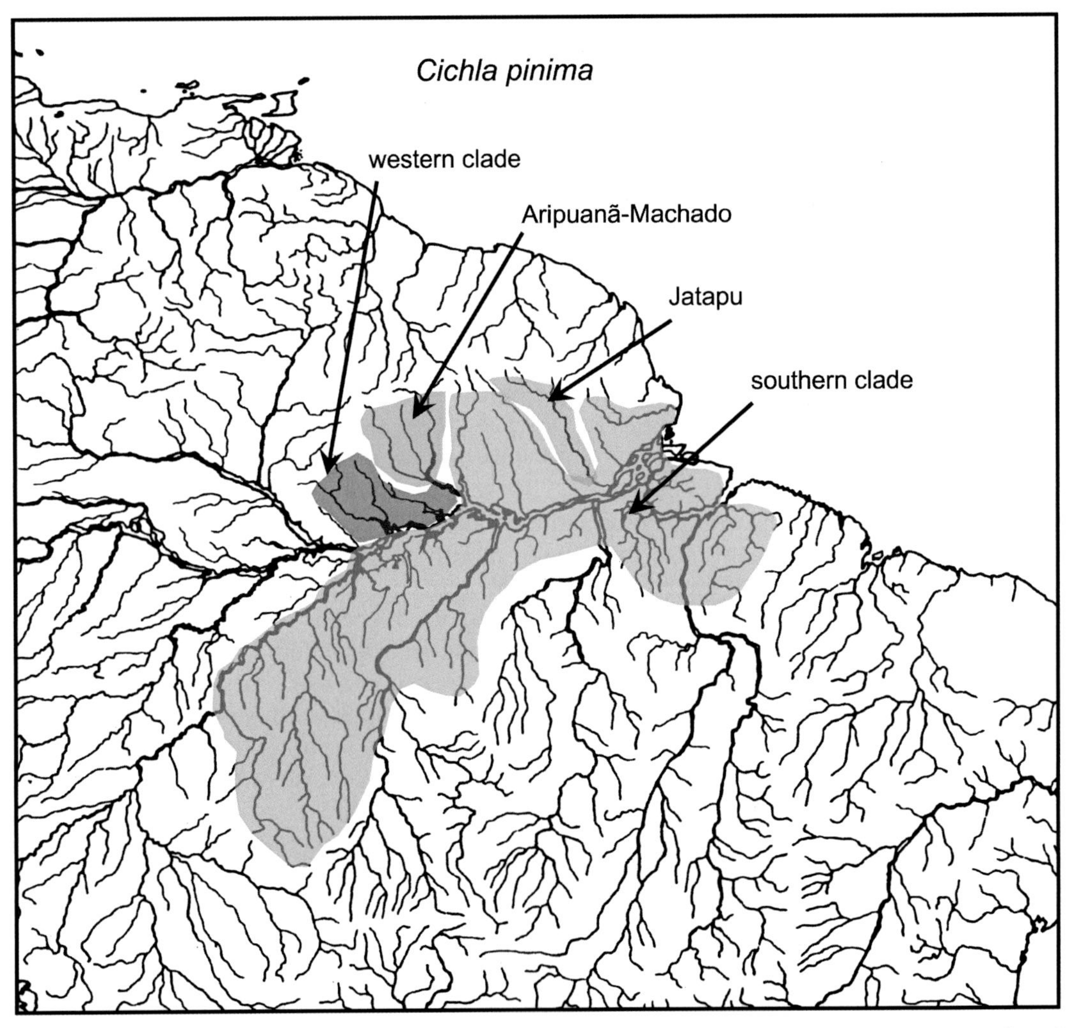

The different colors within the *Cichla pinima* range correspond to genetically differentiated lineages. Map is adapted from a figure in [13].

A similar situation of regional genetic differentiation exists for *C. ocellaris*. Stocks we have identified as "varieties" possess genetic differences that suggest periods of isolation and limited gene flow (i.e., limited interbreeding among stocks). Recent phylogenetic analysis by Willis indicates at least six regional stocks of *C. ocellaris*. However, these stocks also reveal genetic evidence that indicates recent interbreeding (exchange of genes), which supports the idea that *C. ocellaris* is a single species with a broad geographic distribution. Since these stocks appear capable of interbreeding, and apparently have done so naturally at various intervals, we consider them to represent a single species with regional stocks that most likely have adapted to site-specific conditions. Moreover, there are no anatomical or coloration differences that can consistently identify any of these stocks as unique from all others.

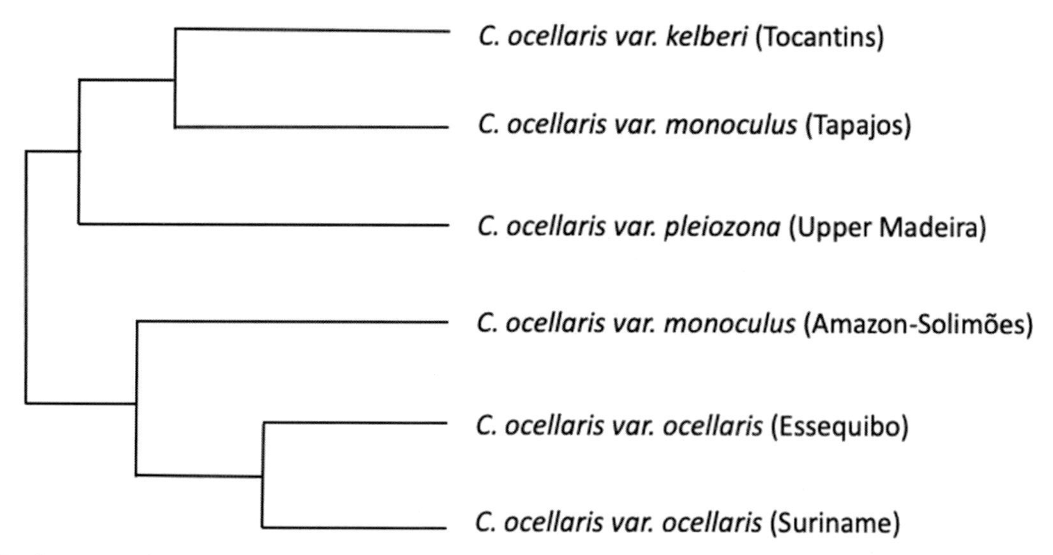

Evolutionary relationships of regional stocks (varieties) of *C. ocellaris* estimated from analysis of DNA sequence data [9].

A brilliant green *C. ocellaris* var. *monoculus* from the Casiquiare River. *Photo: K. Winemiller.*

Cichla intermedia from the Caura River differ genetically (in terms of mitochondrial DNA sequences) when compared to *C. intermedia* from other locations in the Orinoco Basin, but all are currently considered a single species. Similarly, *C. mirianae* from the upper Xingu and upper Tapajos show minor genetic differences, but at this time are considered a single species. Analysis of combined nuclear and mitochondrial DNA of *C. orinocensis* specimens from locations throughout the Orinoco and Negro basins indicates a single species; however, differences in their mitochondrial DNA separate fish populations from the upper and lower reaches of the Rio Negro [9].

Royal peacock bass (*C. intermedia*) from the Manipitare River in the Casiquiare drainage of southern Venezuela. *Photo: K. Winemiller.*

Zoogeography

Peacock bass of one kind or another are found throughout the Amazon and Orinoco River basins, as well as the small rivers draining from the Guyana Shield to the Atlantic Ocean (see map of South American river basins in Chapter 1). Peacock bass do not naturally occur in river drainages on the western slope of the Andes Mountains or coastal drainages of the Caribbean in Colombia and Venezuela (exclusive of the Orinoco) and coastal drainages of eastern and southern Brazil. At least two species, *C. piquiti* and *C. ocellaris* var. *kelberi*, have been introduced into the Paraná-Paraguay Basin in southern Brazil and Paraguay as well as reservoirs on rivers flowing to the Atlantic Ocean in eastern Brazil. *Cichla ocellaris* var. *ocellaris* and perhaps other subspecies and species have been stocked in reservoirs in regions beyond South America, including Panama, Hawaii, Puerto Rico, Malaysia, and Singapore. *Cichla ocellaris* was introduced and thrives in canals within and around the city of Miami, Florida, and the species reportedly has become established in southwest Florida in the canal system between Naples and the Everglades.

The yellow peacock bass, *C. ocellaris* var. *kelberi*, has been introduced extensively in rivers of southern and eastern Brazil where peacock bass do not historically occur. *Photo: K. Winemiller.*

The Amazon and Orinoco river basins are home to an extraordinary diversity of plants and animals, including freshwater fish that number nearly 3000 species at last count [14]. The Casiquiare River connects these basins by flowing from the Upper Orinoco and emptying into the Upper Rio Negro [11, 15, 16]. Despite the existence of this modern corridor for aquatic dispersal, most fish species are found only in one or the other basin, with only about one quarter of the total species documented from the two basins occurring in both [16]. *Cichla temensis*, for example, is broadly distributed throughout both the Orinoco and Rio Negro basins. Other fish, such as discus fish (*Symphysodon* species), silver arowana (*Osteoglossum bicirrhosum*), and South American lungfish (*Lepidosiren paradoxa*), are only found in the Amazon Basin. In several fish genera, the most closely related species (sister taxon) have nonoverlapping distributions, with only one occupying a given basin. For example, the redbelly piranha, *Pygocentrus nattereri*, is found in the Amazon Basin, whereas its sister species, *P. cariba*, only occurs in the Orinoco Basin.

Two species of redbelly piranha, *Pygocentrus nattereri* (top) from the Amazon Basin and *P. cariba* (bottom) from the Orinoco basin. *Photos: K. Winemiller.*

Anglers in search of peacock bass in the Amazon also encounter diverse wildlife. An emerald tree boa (*Corallus caninus*) in the riparian forest of the lower Rio Negro. *Photo: K. Winemiller.*

Orinoco Basin

The Orinoco River is the second largest river in South America according to discharge (average annual discharge is 37,000 m^3 per second) and third largest in terms of area (830,000 km^2) [17]. The Orinoco source is located in the Parima Mountains on the border between Venezuela and Brazil, and from there flows northwest and then northeast to the Caribbean Gulf of Paria. During the rainy season, the Orinoco carries large sediment loads and nutrients from Andean tributaries, such as the Guaviare, Meta, Apure, and Arauca. During the rainy season, thousands of square kilometers of savanna and gallery forest are flooded in the Llanos region of Colombia and Venezuela, which provides important habitat and food resources for peacock bass and most other fish. In the western Llanos, the Capanaparo, Cinaruco, Bita, and Tomo rivers have clear, moderately acidic waters with low concentrations of suspended sediments. These rivers provide excellent habitat for peacock bass, as do clearwater rivers, such as the Aguaro, that flow southward from the Venezuelan coastal mountains to the lower Orinoco.

Three species of peacock bass (*C. intermedia*, *C. orinocensis*, and *C. temensis*) can be found near rocky shoals in the Cinaruco River, Venezuela. *Photo: K. Winemiller.*

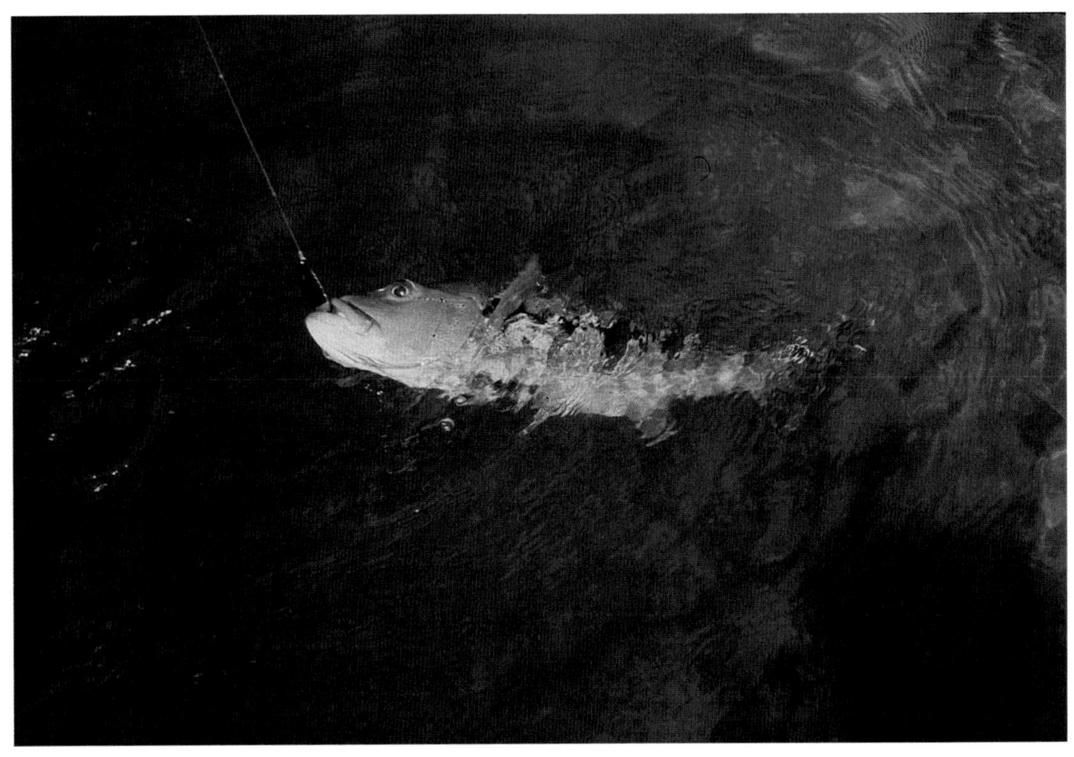

A royal peacock bass (*C. intermedia*) fighting in the clear water of the Cinaruco River in the Venezuelan Llanos. *Photo: K. Winemiller.*

Major blackwater tributaries of the Orinoco draining the ancient, weathered landscapes of the Guiana Shield include the Aro, Caura, and Caroní [18]. These rivers provide suitable conditions for peacock bass. The Caroní River historically did not have peacock bass in reaches above cataracts on its lower reaches, but with the construction of Guri Reservoir, peacock bass were introduced into the upper basin. The Upper Orinoco is a mosaic of blackwater (Atabapo, Inirida) and clearwater (Ventuari, Mavaca, Ocamo) rivers, all of which support large populations of multiple peacock bass species. The extreme black waters of the Atabapo River join the Orinoco near the town San Fernando de Atabapo. The Guaviare from Colombia also joins the Orinoco at this location, and the contrast of waters is very apparent, with clear water entering from the Upper Orinoco and Ventuari, black water entering from the Atabapo and Inirida, and white water entering from the Guaviare, all of which mix together as they flow downstream in the Orinoco. The Casiquiare River, which links the Upper Orinoco and Rio Negro, has mostly clearwater tributaries (e.g., Pamoni, Siapa) in its upper reaches (nearer to the Orinoco), and extreme blackwater tributaries (e.g., Pasimoni) in its lower reaches (nearer to the Rio Negro) [15]. The largest number of naturally occurring peacock bass species can be found together (*C. intermedia, C. ocellaris* var. *monoculus, C. orinocensis, C. temensis*) in the Casiquiare and Upper Orinoco.

Three peacock bass species (top—*C. orinocensis*, middle—*C. ocellaris* var. *monoculus*, bottom—*C. intermedia*) captured from the same habitat in the Manipitare River (tributary of the Siapa River, a tributary of the Casiquiare River). Not pictured here is *C. temensis*, a fourth species found in the Manipitare River. *Photo: K. Winemiller.*

Cichla orinocensis and *C. temensis* are found in clearwater and blackwater habitats throughout the Orinoco Basin as well as the Rio Negro, whereas *C. intermedia* seems to be restricted to clearwater rivers of the Orinoco Basin, and *C. ocellaris* var. *monoculus* is limited to the Casiquiare and Upper Orinoco. It is unclear why *C. ocellaris* var. *monoculus* has not invaded blackwater and clearwater habitats further downstream in the Orinoco, but competition with *C. orinocensis* in these regions might be a factor. The failure of *C. intermedia* to invade the lower Casiquiare and Rio Negro appears to be due to its intolerance of extreme blackwater conditions. *Cichla orinocensis* and *C. temensis* apparently have not dispersed more widely in the Amazon River due to their intolerance of the muddy whitewater conditions of the Solimões-Amazon, although *C. temensis* somehow did manage to cross this ecological barrier to colonize suitable habitats within tributaries entering the lowest reaches Madeira River.

The speckled peacock bass (*C. temensis*) is distributed in clearwater and blackwater rivers throughout the Orinoco and Negro river basins. *Photo: C. Montaña.*

Amazon River

The Amazon is the world's largest river, both in terms of basin area (7,050,000 km^2) and discharge (about 6590 km^3 per year, comprising about 20% of the global freshwater discharge), and it also is the second longest river, spanning 6400 km from its headwaters in the Andes to its mouth on the Atlantic coast. The Amazon has nine major tributaries. The Ucayali, Marañon, and Napo rivers drain the Andes in Peru and Ecuador. The Caquetá River flows from Colombia to Brazil where it is called the Japurá River, and the Putumayo flows from Colombia and Ecuador to Brazil where it is called the Içа River. To the south, the Madeira River has its headwaters in the Andes, flowing from Bolivia, southern Peru, and southern Brazil to join the Amazon a short distance downstream from the city of Manaus. All of the major tributaries of the western Amazon are whitewater rivers, carrying high sediment loads during the rainy season, with neutral pH and more nutrients as compared to blackwater rivers of the Guiana Shield, such as the Negro and Trombetas, and clearwater rivers of the Brazilian Shield, such as the Tapajos, Xingu, and Tocantins [19].

The muddy waters of the Amazon River are unsuitable for peacock bass. *Photo: K. Winemiller.*

An important ecosystem in the Amazon ecosystems is the várzea, seasonally flooded terrain along the course of the Amazon and lower reaches of its tributaries. These floodplains cover more than 100,000 km^2 [20] and contain a rich mosaic of aquatic and seasonally flooded terrestrial habitats that support high fish diversity and important fisheries [21, 22]. From the western border of Brazil (at the town of Tabatinga, Brazil) to its confluence with the Rio Negro, the Amazon is called the Solimões River. The Solimões carries high amounts of suspended clays and other fine particles eroded from the Andes Mountains. Some lakes and channels in the Solimões floodplains have black waters that support *Cichla ocellaris* var. *monoculus*, the only species of peacock bass that appears to occur naturally in the vast lowlands of the western Amazon [7, 8]. When the Rio Negro's black waters join the white waters of the Solimões at the "great meeting of the waters" near the city of Manaus, the two waters mix only gradually over the next hundred kilometers, this separation owing to differences in temperature, density, and chemistry.

Lakes in the Amazon várzea (floodplain) sometimes have clear water suitable for *Cichla ocellaris* var. *monoculus*. *Photo: K. Winemiller.*

On the northern side of the Amazon River, several major clearwater and blackwater tributaries drain the Guiana Shield. These include the Preto da Eva, Uatumã, Trombetas, Jari, Paru, and Araguari. The clearwater Trombetas flows from the Tumucumaque highlands and has many cataracts along its course, some of which were submerged under the huge reservoir formed by the Balbina Hydroelectric Dam. The lower reaches of the Trombetas have broad floodplains with many natural lakes that provide habitat for peacock bass. The Araguari River is located near the Amazon mouth and is not really a tributary but flows directly to the Atlantic. Several large tributaries are located on the southern side of the Amazon River. The Madeira is a whitewater river that flows from the Andes Mountains and Bolivian Alto Plano to join the Amazon downstream from Manaus. Major clearwater tributaries draining the Brazilian Shield are the Tapajos, Xingu, and Araguaia-Tocantins. These southern tributaries are discussed separately later.

When fishing in the Amazon, anglers sometimes catch the acrobatic arowana (*Osteoglossum bicirrhosum*) on the same lures used to fish for peacock bass. *Photo: L. Kelso Winemiller.*

The muddy waters of the Amazon limit dispersal by clearwater and blackwater-adapted fish, such as peacock bass. Only species fairly tolerant of whitewater conditions can migrate between tributaries of the Brazilian Shield in the south and those of the Guiana Shield to the north [10, 15, 23, 24]. Apparently, *C. ocellaris* var. *monoculus* and *C. pinima* are two peacock bass species that have limited capability to disperse along the course of the lower Amazon. In the lower Amazon, *Cichla temensis*, *C. orinocensis*, and *C. ocellaris* var. *monoculus* occur in the Rio Preto da Eva, a small blackwater tributary located approximately 60km east from Manaus. *Cichla temensis* has also been reported from the lower Uatumã (Jatapu) River, whereas *C. ocellaris* var. *monoculus* and *C. pinima* have been recorded from the middle to upper Uatumã [7, 8] as well as the Paru, Jari, and Araguari rivers. The Paru and Jari rivers have moderate black waters (pH5–6), whereas the Araguari River is a clearwater river that flows over silt and clay deposited from the Amazon River. *Cichla pinima* is found in the Jari River above and below the Cachoeira de Santo Antonio [8, 13].

Madeira River

The Madeira Basin covers close to a quarter of the Brazilian Amazon, and stretches over 1.5 million square kilometers across Peru, Bolivia, and Brazil. The second largest Amazon tributary according to discharge, the Madeira is formed below the confluences of the Beni, Madre de Dios, Mamoré, and Guaporé rivers that originate in the Andes Mountains and Bolivian-Peruvian Alto Plano (high plains). This colossal whitewater river carries heavy loads of sediments eroded from the Andes. Tributaries that enter the middle and lower Madeira from the Brazilian Shield region in the east have clear or black waters with low concentrations of dissolved nutrients. These eastern tributaries contain suitable habitat for several peacock bass species, including *C. ocellaris* var. *monoculus, C. pinima*, and *C. temensis* (the latter only found in blackwater habitats in tributaries of the lower Madeira). In addition to large rivers such as the Jiparaná (also called Machado River), Taya, Maués Açu, Abacaxis, and Aripuanã, many small rivers, streams, and floodplain lakes along the lower Madeira support peacock bass [7, 8]. *Cichla ocellaris* var. *pleiozona* is the only peacock bass that naturally occurs in the upper Madeira Basin, the region above the São Antonio Falls, which now are submerged beneath the reservoir formed by the Santo Antônio Hydroelectric Dam near the city of Porto Velho.

Flooded forest provides essential habitat for many fish species of the Amazon region, including peacock bass. *Photo: K. Winemiller.*

Rio Negro

Rio Negro is the largest blackwater river in the world and third largest tributary of the Amazon by discharge. The upper course of the Rio Negro in Colombia is called the Guainía River, which flows east then south to form the border between Colombia and Venezuela before entering northern Brazil. Blackwater tributaries include the Içana, Uaupés (called the Vaupés in Colombia), Marie, Demini, and Unini rivers. The Branco River is a clearwater tributary (sometimes characterized as white water, but suspended sediment loads are not nearly has high as those of the Solimões-Amazon) that flows from the border of Brazil with Guyana through the savannas of Brazil's Roraima state before joining the Negro. The Rio Negro is extremely nutrient poor with low pH (4–5.5) caused by high concentrations of dissolved organic compounds leached from submerged leaves and wood [25]. Over nearly its entire course, the Rio Negro has a very low elevational gradient and broad floodplains, which creates environmental conditions highly favorable for peacock bass. The annual flood pulse is high (several meters in most regions) and sustained (high water from February to September). The Rio Negro joins the Amazon at the "great meeting of waters" near Manaus.

The extreme black waters of the Rio Negro are strongly acidic and nutrient poor, and yet the largest peacock bass are found in this system. *Photo: K. Winemiller.*

Rio Negro has exceptionally high fish diversity (389 genera and at least 1165 species [26]) that includes several migratory species, most notably the jaraqui (*Semaprochilodus insignis*) that annually migrates from the Rio Negro to the Amazon for spawning and exploiting productive floodplain habitats as nursery grounds [27]. More than 150 fish species are considered endemic to the Rio Negro Basin, including four genera of tetras (*Tucanoichthys, Ptychocharax, Atopomesus, Leptobrycon* [Characidae]) [26].

Several kinds of stingrays are common in the Rio Negro. This spotted one is *Potamotrygon motoro*. *Photo: K. Winemiller.*

Cichla temensis, C. orinocensis, and *C. ocellaris* var. *monoculus* occur in channels and lakes throughout the Rio Negro Basin as well as the Casiquiare River that enters the headwaters from the Upper Orinoco. The Rio Negro supports the finest sport fisheries for speckled and butterfly peacock bass, with many angling world records recorded from remote and lightly exploited areas in the middle and upper portions of the basin. Other important areas for sport fishing are in blackwater tributaries of the lower Branco River, such as the Tapera and Xeriuini rivers [20]. Previously discussed were the enigmatic distribution of *C. ocellaris* var. *monoculus* in the lower Branco and *C. ocellaris* var. *ocellaris* in the upper Branco, and the existence of genetic differences between stocks of *C. orinocensis* inhabiting the upper versus middle/lower Rio Negro.

Juvenile *C. temensis* captured using a seine net near the shoreline of the Rio Negro. *Photo: K. Winemiller.*

Speckled peacock bass (*C. temensis*) thrive in the black waters of the Rio Negro. *Photo: K. Winemiller.*

Tapajos River

The Tapajós is a major clearwater river of the Brazilian Shield that is formed by the confluence of the Juruena and Teles Pires rivers. From this confluence, the Tapajos flows approximately 780km to the Amazon at the city of Santarém. The river has a series of cataracts and rapids in its reach between the Juruena River and the Maranhão Grande Cataracts. In the final 160km, the river is very wide (6.4–14.5km) and deep (25m). Along its lower course, the Tapajós valley is bordered on both sides by bluffs ranging from 90 to 120m high. The Tapajos carries few suspended sediments compared to whitewater rivers and low concentrations of dissolved organic compounds compared to blackwater rivers. The Tapajos also has higher mineral content than most black waters, and a pH ranging from slightly acid to basic. The Tapajos like all other Amazon River drainages has a species-rich freshwater fauna with Characiformes and Siluriformes being the dominant orders. Characidae (Characiformes), Cichlidae (Perciformes), and Loricariidae (Siluriformes) are the most species-rich families.

Clear water of the lower Tapajos River. *Photo: S. Willis.*

Cichla pinima is common in tributaries of the middle Tapajos River in the vicinity of the towns of Itaituba and Jacareacanga and throughout the lower Tapajos to its mouth at Santarém. *Cichla ocellaris* var. *monoculus* often co-occurs with *C. pinima* in these middle and lower reaches of the Tapajos. *Cichla pinima* also is captured in suitable habitats in the upper Tapajos, including the lower reaches of the Juruena and Teles Pires rivers. *Cichla mirianae* is the only peacock bass found in the upper

Tapajos within the Juruena, Teles Pires, and São Benedito rivers where they are targeted by sport fishers. In this region, anglers exploit other abundant game fish, including peche cachorro (*Hydrolycus scomberoides*), jau (*Zungaro zungaro*), matrinxãs (*Brycon* species), and pacu (*Myleus* species).

The city of Santarem is located where the Tapajos River joins the Amazon. *Photo: K. Winemiller.*

Xingu River

The Xingu is a major clearwater river that flows from the forests and savannas of Brazil's Mato Grosso state northward nearly 2000 km to join the Amazon River. Before emptying into the Amazon just downstream from the Ilha Grande de Gurupá, the Xingu channel expands into an enormous lake wherein its waters mix with those of the Amazon entering through natural channels that wind through a forested archipelago. The Xingu's headwaters include the Curiseu, Culuene, Batovi, Suia Missu, and Romuro rivers. In the river's lower reaches,

Fire peacock bass (*C. mirianae*) from the São Benedito. *Photo: K. Winemiller.*

Sloths are often observed in the trees along waterways in the Amazon. *Photo: K. Winemiller.*

the Iriri River is the main tributary that joins several kilometers upstream from the town of Altamira do Xingu. The Iriri, Xingu, and most of its headwater tributaries have clear waters characteristic of the ancient weathered landscape of the Brazilian Shield. The upper basin has relatively flat topography, and the middle portion of the Xingu (region above and below the Iriri confluence and Altamira do Xingu) has a higher elevational gradient as the river descends from the Brazilian Shield formation to the Amazon lowlands. This middle reach contains a mosaic of channels and rocky islands and numerous rapids that provide habitat for dozens of endemic fishes adapted to living in fast water. Among these endemic fishes is *Cichla melaniae*, which was abundant in swiftly flowing and rocky channels in the Volta Grande reach of the middle Xingu prior to construction of the Pimental Dam (part of the Belo Monte Hydroelectric Project) that converted a long stretch of the river into a reservoir. The rapids complexes in Volta Grande region formed a barrier to upstream dispersal by lowland-adapted fish, such as *Cichla pinima* and *C. ocellaris* var. *monoculus*. Consequently, these two species are found only in the lower Xingu River downstream from the rapids. *Cichla mirianae* is found in the upper Xingu and its tributaries, including the Batovi, Culuene, Fresco, and Suia Missu rivers, where it is targeted by anglers.

The swift, clear waters of the Xingu River are home to *Cichla melaniae*. *Photo: K. Winemiller.*

The rapids region of the lower Xingu River is home to dozens of species of loricariid catfish, commonly called "plecos" in the aquarium hobby. Many plecos are endemic to the Xingu rapids (occurring nowhere else). This is *Panaque armbrusteri*, a wood-eating pleco. *Photo: K. Winemiller.*

Retroculus xinguensis, a cichlid endemic to the swiftly flowing waters of the Xingu River. *Photo: H. López-Fernández.*

Araguaia-Tocantins Basin

The Tocantins is a clearwater river that empties into the Bahia de Marajó which also receives a small fraction of the outflow from the Amazon through channels connecting to the main estuary located north of Ilha de Marajó. The Tocantins is formed by the Maranhão and Paranã rivers on the eastern edge of the Brazilian Shield and flows northwards over a different sedimentary basin to join the Bahia upstream from the city of Belém. The Araguaia River, the principal tributary of the Tocantins, arises in the Caiapo Mountains and flows parallel to the Tocantins. Unlike the Tocantins, the Araguaia has a shallow elevational gradient with a sinuous channel, turbid water, and a broad floodplain containing extensive wetlands. The Araguaia has the largest fluvial island in the world, Ilha do Bananal [28]. Rapids and cataracts are found in the upper reaches of both rivers, but most of the rapids in the lower Tocantins were flooded when the Tucuruí Hydroelectric Dam was constructed. The rapids on the lower reaches, and now the dam, were barriers to upstream dispersal by lowland-adapted Amazon fishes.

The fish fauna for the Araguaia-Tocantins consists of approximately 300 species [29], with many of them endemic to the basin. These include several headstanders (Anostomidae, e.g., *Laemoloita petiti*, *Leporinus affinis*, *Sartor tucuruiense*), a piranha (*Serrasalmus geryi*), a pacu (*Mylesinus pauscisquamatus*), a driftwood catfish (*Tocantinsia depressa*), and pike cichlids, such as the dwarf species *Crenicichla compressiceps*, that inhabit rapids in the lower Tocantins [30].

The many lakes in the Araguaia floodplains provide excellent habitat for peacock bass. *Cichla piquiti* (Tucunare azul) and *C. ocellaris* var. *kelberi* are the only peacock bass native to the Araguaia-Tocantins Basin above the rapids on the lower Tocantins. In the lower Tocantins, both species are found in the Itacaiunas, Marabá, and Itupiranga rivers and are common in main channel as well as the floodplain lakes of the Araguaia River. *Cichla pinima* is common in the lowest reaches of Tocantins River, where it sometimes co-occurs with *C. piquiti* and *C. ocellaris. Cichla piquiti and C. ocellaris* var. *kelberi* from the Araguaia-Tocantins have been widely introduced into reservoirs of the Upper Paraná River drainage in the Brazilian States of Minas Gerais, São Paulo, and Paraná [31].

The blue peacock bass (*C. piquiti*) has a native distribution restricted to the Araguaia-Tocantins Basin. This species adapts well to conditions in ponds and reservoirs and consequently has been introduced widely in other regions of Brazil. *Photo: K. Winemiller.*

Guiana Shield Coastal Rivers

Cichla ocellaris, the first peacock bass collected by early naturalists, is native to the Guiana Shield region in present-day Guyana, Suriname, and French Guiana. The eastern limit of its distribution is the Marowijne River and the western limit appears to be the Takutu River (upper Branco) [7, 32, 33]. *Cichla ocellaris* var. *ocellaris* (pond lukunani) and the falls lukunani (*C. cataractae*) both are found in the Essequibo Basin. The largest river in the region, the Essequibo has two major tributaries, the Cuyuni and Mazaruni, that join its lower course near the town of Bartica. The Cuyuni originates in eastern Venezuela, and the Mazaruni flows from the Pakaraima Mountains in northwest Guyana.

The Rewa River flows to the Rupununi River, a tributary of the Essequibo in Guyana. *Photo: D. Taphorn.*

Cichla temensis was reported to occur in the Cuyuni River near El Dorado, Venezuela [34], but we have surveyed fish in this region and cannot verify that claim. Previous surveys conducted near Isla Anacoco in the lower Cuyuni River (near San Martin of Turumban, Venezuela) produced only *C. ocellaris* [10]; however, one specimen was later determined to be the falls lukunani. *Cichla ocellaris* is common in the Essequibo River and its tributaries, including the Rupununi, Burro-Burro, and Siparuni [35], where it has long sustained sport and subsistence fisheries. In Suriname, *Cichla ocellaris* is found in the Suriname, Saramacca, Nickerie, and Marowijne (Maroni) rivers, the latter bordering French Guiana [32]. Fish from the Suriname River and Brokopondo Reservoir are marketed in the aquarium hobby as "Brokopondo peacock bass," which have been shown to possess some genetic differences from *C. ocellaris* var. *ocellaris* from Guyana.

Cichla ocellaris var. *ocellaris* is common in coastal drainages in Guyana, Suriname, and French Guiana. *Photo: D. Taphorn.*

The Rupununi River in Guyana. *Photo: D. Bloom.*

According to Stuart Willis, *C. ocellaris* from tributaries of the upper Branco (Pirara and Takutu rivers) are closer genetically to *C. ocellaris* var. *ocellaris* from the Essequibo than *C. ocellaris* var. *monoculus* from the Rio Negro, which indicates considerable dispersal between the upper Branco (Amazon Basin) and upper Essequibo within the Rupununi Savanna District of Guyana. This region floods extensively during the rainy season, and fish dispersal between the adjacent drainages was initially proposed by the pioneering fish ecologist Rosemary Lowe-McConnell over 50 years ago [35].

Hybridization

As previously mentioned, peacock bass species are notoriously difficult to identify, and some experts believe that some of the fish considered to be distinct and undescribed species are in fact naturally occurring hybrids between different *Cichla* species. Evidence of hybridization between *Cichla* species has been observed in natural populations of the Amazon region [36, 37] and among nonnative stocks introduced into reservoirs and ponds both within and outside the historical geographic ranges of the species.

Evidence of historical gene exchange, including ancient hybridization events, is not uncommon even for *Cichla* species that seem to be quite distinct and unlikely to interbreed under natural conditions. Examples include *C. ocellaris* var. *monoculus* with *C. temensis* from the Mavaca River (Upper Orinoco) and *C. intermedia* with *C. orinocensis* from the Parguaza River (Orinoco tributary) [8]. There also is some evidence of natural hybridization between *C. ocellaris* var. *monoculus* and *C. temensis* from the Central Amazon [36, 37]. In the lower Rio Negro, *Cichla orinocensis* and *C. ocellaris* var. *monoculus* share genetic patterns indicative of historic interbreeding [9]. Interestingly, fish from the upper Rio Negro (above the mouth of the Rio Branco) do not show signs of hybridization. Shared gene sequences between *C. pinima* from the lower Tocantins and *C. piquiti* from the Araguaia-Tocantins upstream suggest at least some degree of interbreeding occurred between ancestral lineages of these species. More recent interbreeding between these two species appears to have occurred in the lower Tocantins below the Tucuruí Dam, possibly caused by ecological changes brought about by an altered flow regime [8].

Hybridization apparently is more likely to occur when peacock bass species are introduced into regions where they are not native, or two or more species are stocked into bodies of water that were created or altered by humans. There is genetic evidence of hybridization between *C. temensis* and *C. orinocensis* from Guri Reservoir on the Caroní River, an Orinoco tributary where neither of these species apparently had occurred naturally [8]. Some of these specimens were previously misidentified as either *C. ocellaris* or *C. intermedia*. Hybridization between *C. piquiti* and *C. ocellaris* var. *kelberi* stocked into reservoirs of the Upper Paraná Basin has been reported by multiple studies [38–40]. These two species escaped from reservoirs and established self-sustaining stocks in the Paraguay and Paraná rivers where they also have hybridized [38].

A peacock bass from Guri Reservoir, Venezuela, that has coloration that is superficially similar to *C. ocellaris*, but actually is a hybrid between *C. orinocensis* and *C. temensis*. *Photo: K. Winemiller.*

Black waters of Guri Reservoir on the Caroní River, Venezuela. *Photo: K. Winemiller.*

A peacock bass from Guri Reservoir near the inflow of the Chiuao River. This fish could be a hybrid between *C. orinocensis* and *C. temensis* or possibly an undescribed form that was undetected in the Upper Caroní River prior to construction of the huge reservoir. *Photo: Daniel González.*

Peacock bass caught from an artificial pond near Belém in Pará state, Brazil. This fish appears to be a hybrid, possibly between *C. pinima* and *C. ocellaris* or *C. piquiti* (note blue coloration of the fins and lack of black spots in the postorbital region of the head). *Photo: K. Winemiller.*

References

[1] S.O. Kullander, Cichlid Fishes of the Amazon River Drainage of Peru, Swedish Museum of Natural History, Stockholm, 1986.

[2] M.L.J. Stiassny, The phyletic status of the family Cichlidae (Pisces: Perciformes): a comparative anatomical investigation, Neth. J. Zool. 31 (1981) 275–314.

[3] M.L.J. Stiassny, Cichlid familial intrarelationships and the placement of the neotropical genus *Cichla* (Perciformes, Labroidei), J. Nat. Hist. 21 (1987) 1311–1331.

[4] M.L. Stiassny, Phylogeny intrarelationships of the family Cichlidae: an overview, in: M.H. Keenlyside (Ed.), Cichlid Fishes: Behaviour, Ecology and Evolution, Chapman Hall, London, 1991.

[5] H. López-Fernández, K.O. Winemiller, R.L. Honeycutt, Multilocus phylogeny and rapid radiation in Neotropical cichlid fishes (Perciformes: Cichlidae: Cichlinae), Mol. Phylogenet. Evol. 55 (2010) 1070–1086.

[6] H. Lopez-Fernandez, J.H. Arbour, K.O. Winemiller, R.L. Honeycutt, Testing for ancient adaptive radiations in Neotropical cichlid fishes, Evolution 67 (2013) 1321–1337.

[7] S.O. Kullander, E.J.G. Ferreira, A review of the South American cichlid genus *Cichla*, with descriptions of nine new species (Teleostei: Cichlidae), Ichthyol. Explor. Freshw. 17 (2006) 289–398.

[8] S.C. Willis, J. Macrander, I.P. Farias, G. Orti, Simultaneous delimitation of species and quantification of interspecific hybridization in Amazonian peacock cichlids (genus *Cichla*) using multi-locus data, BMC Evol. Biol. 12 (2012) 96.

[9] S.C. Willis, I.P. Farias, G. Orti, Multi-locus species tree for the Amazonian peacock basses (Cichlidae: *Cichla*): emergent phylogenetic signal despite limited nuclear variation, Mol. Phylogenet. Evol. 69 (3) (2013) 479–490.

[10] S.C. Willis, M.S. Nunes, C.G. Montana, I.P. Farias, N.R. Lovejoy, Systematics, biogeography, and evolution of the neotropical peacock basses *Cichla* (Perciformes: Cichlidae), Mol. Phylogenet. Evol. 44 (1) (2007) 291–307.

[11] S.C. Willis, M. Nunes, C.G. Montaña, I.P. Farias, G. Ortí, N. Lovejoy, The Casiquiare River acts as a dispersal corridor between the Amazonas and Orinoco River basins: biogeographic analysis of the genus *Cichla*, Mol. Ecol. 19 (2010) 1014–1030.

[12] S.C. Willis, K.O. Winemiller, C.G. Montaña, J. Macrander, P. Reiss, I.P. Farias, G. Ortí, Population genetics of the speckled peacock bass (*Cichla temensis*), South America's most important inland sport fishery, Conserv. Genet. 16 (2015) 1345–1357.

[13] S.C. Willis, One species or four? yes! or, arbitrary assignment of lineages to species obscures the diversification processes of Neotropical fishes, PLoS One 12 (2) (2017), e0172349.

[14] J.S. Albert, R.E. Reis, Historical Biogeography of Neotropical Freshwater Fishes, University of California Press, Berkeley, 2011.

[15] K.O. Winemiller, H. López Fernández, D.C. Taphorn, L.G. Nico, A. Barbarino Duque, Fish assemblages of the Casiquiare River, a corridor and zoogeographic filter for dispersal between the Orinoco and Amazon basins, J. Biogeogr. 35 (2008) 1551–1563.

[16] K.O. Winemiller, S.C. Willis, Biogeography of the Casiquiare river and Vaupes arch, in: J.S. Albert, R.E. Reis (Eds.), Historical Biogeography of Neotropical Fishes, Verlag Dr. Friedrich Pfeil Scientific Publishers, Munich, Germany, 2011, pp. 225–242 (Chapter 11).

[17] J.G. Lundberg, L.G. Marshall, J. Guerrero, B. Horton, M.C.L. Malabarba, F. Wesselingh, The stage for Neotropical fish diversification: a history of tropical South American Rivers, in: L. Malabarba, R.E. Reis, R.P. Vari, Z.M.S. Lucena, C.A.S. Lucena (Eds.), Phylogeny and Classification of Neotropical Fishes, Museu de Ciências e Tecnologia, Porto Alegre, 1998.
[18] N. Hubert, J.F. Renno, Historical biogeography of South American freshwater fishes, J. Biogeogr. 33 (2006) 1414–1436.
[19] H. Sioli, The Amazon and its main afluents: hydrography, morphology of the river courses, and river types, in: H. Sioli (Ed.), The Amazon, Monographie Biologicae, Dr. W. Junk Publishers, The Hague, Netherlands, 1984, p. 56.
[20] M. Goulding, R. Barthem, E. Ferreira, R. Duenas, The Smithsonian Atlas of the Amazon, Smithsonian Books, Washington, DC, 2003.
[21] W.J. Junk, M.G.M. Soares, U. Saint-Paul, The fish, in: W.J. Junk (Ed.), The Central-Amazonian Floodplain: Ecology of a Pulsing System, Ecological Studies, Springer Verlag, Berlin, 1997.
[22] C.C. Arantes, K.O. Winemiller, M. Petrere, L. Castello, L.L. Hess, C.E. Freitas, Relationships between forest cover and fish diversity in the Amazon River floodplain, J. Appl. Ecol. 55 (2018) 386–395.
[23] N.R. Lovejoy, M.L.G. de Araújo, Molecular systematics, biogeography and population structure of Neotropical freshwater needlefishes of the genus *Potamorrhaphis*, Mol. Ecol. 9 (2000) 259–268.
[24] J.F. Renno, N. Hubert, J.P. Torrico, F. Duponchelle, J. Nunez Rodriguez, C. Garcia Davila, S.C. Willis, E. Desmarais, Phylogeography of *Cichla* (Cichlidae) in the upper Madeira basin (Bolivian Amazon), Mol. Phylogenet. Evol. 41 (2006) 503–510.
[25] M. Goulding, M.L. Carvalho, E.G. Ferreira, Rio Negro, Rich Life in Poor Water: Amazonian Diversity and Food Chain Ecology as Seen Through Fish Communities, SPB Academic Publishing, The Hague, 1988.
[26] H. Beltrão, J. Zuanon, E. Ferreira, Checklist of the ichthyofauna of the Rio Negro basin in the Brazilian Amazon, ZooKeys 881 (2019) 53–89.
[27] M.C.L.B. Ribeiro, Natural hybrid between two tropical fishes: *Semaprochilodus insignius* vs *Semaprochilodus taeniurus* (Teleostei, Caracoidei, Prochilodontidae), Rev. Bras. Zool. 2 (1985) 419–421.
[28] A.C. Ribeiro, Tectonic history and the biogeography of the freshwater fishes from the coastal drainages of eastern Brazil: an example of faunal evolution associated with a divergent continental margin, Neotrop. Ichthyol. 4 (2006) 225–246.
[29] G.M. Santos, M. Jegu, B. Merona, Catálogo de Peixes Comerciais do Baixo Rio Tocantins, ProjetoTucurui, ELETRONORTE/CNPq/INPA, Manaus, Brasil, 1984.
[30] A. Ploeg, The cichlid genus *Crenicichla* from the Tocantins River, state of Pará, Brazil, with descriptions of four new species (Pisces, Perciformes, Cichlidae), Beaufortia 36 (1986) 57–80.
[31] T. Ferraz-Luiz, M. Roquetti Velludo, A. Carvalho Peret, J.L. Rodrigues Filho, A. Moldenhauer Peret, Diet, reproduction and population structure of the introduced Amazonian fish *Cichla piquiti* (Perciformes: Cichlidae) in the Cachoeira Dourada reservoir (Paranaíba River, Central Brazil), Rev. Biol. Trop. 59 (2) (2011) 727–741.
[32] S.O. Kullander, H. Nijssen, The Cichlids of Suriname, Brill, Leiden, The Netherlands, 1989.
[33] S.O. Kullander, Cichlidae (Cichlids), in: R.E. Reis, S.O. Kullander, C.J. Ferraris Jr. (Eds.), Checklist of the Freshwater Fishes of South and Central America, Edipucrs, Porto Alegre, 2003.
[34] A. Machado-Allison, B. Chernoff, R. Royer, F. Mago Lecia, J. Velásquez, C. Lasso, H. López-Rojas, A. Bonilla, F. Provenzano, C. Silvera, Ictiofauna de la cuenca del río Cuyuní en venezuela, Interciencia 25 (1) (2000) 13–21.
[35] R.H. Lowe-McConnell, The cichlid fishes of Guyana, South America, with notes on their ecology and breeding behavior, Zool. J. Linnean Soc. 48 (1969) 255–302.
[36] M.N.A. Brinn, J.I.R. Porto, E. Feldberg, Karyological evidence for interspecific hybridization between *Cichla monoculus* and *C. temensis* (Perciformes, Cichlidae) in the Amazon, Hereditas 41 (2004) 252–257.
[37] A.S. Teixeira, S.S. de Oliveira, Evidence for a natural hybrid of peacock bass (*Cichla monoculus* vs. *Cichla temensis*) based on esterase electrophoretic patterns, Genet. Mol. Res. 4 (1) (2005) 74–83.
[38] G.C. Almeida-Ferreira, Spar genetic analysis of two invasive species of *Cichla* (Tucunaré) (Perciformes: Cichlidae) in the Paraná River Basin, Acta Sci. 33 (1) (2011) 79–85.
[39] A.V. Oliveira, A.J. Prioli, S.M.A.P. Prioli, T.S. Bignotto, H.F. Júlio Junior, H. Carrer, C.S. Agostinho, L.M. Prioli, Genetic diversity of invasive and native *Cichla* (Pisces, Perciformes) populations in Brazil with evidence of interspecific hybridization, J. Fish Biol. 69B (2006) 260–270.
[40] L.S. Gasques, T.M.C. Fabrin, D.D. Gonçalves, S.M.A.P. Prioli, A.J. Prioli, Prospecting molecular markers to distinguish *Cichla kelberi*, *C. monoculus* and *C. piquiti*, Acta Sci. Biol. Sci. 37 (2015) 455–462.

CHAPTER

12

Fisheries, captive care, and conservation

Photo: C. Montaña.

Surrounded by sweltering heat and dripping with sweat, we lugged our aluminum boat across a narrow strip of land separating a floodplain lake from the main channel of the Cinaruco River. It was January 2003, dry season in the Llanos, and we were assisting our colleague, Craig Layman, with his research on the effects of illegal fishing on fish stocks in the Santos Luzardo National Park. Venezuela's President Jaime Lusinchi created the national park, also known as the Cinaruco-Capanaparo National Park, by decree in 1988, with a principal goal being conservation of the aquatic ecosystem supporting one of South America's finest peacock bass fisheries. When we first visited the Cinaruco River in the late 1980s,

https://doi.org/10.1016/B978-0-323-85157-2.00005-6

we were overwhelmed by the abundance of river's three peacock bass species (*C. temensis, C. orinocensis, C. intermedia*). For the greater part of the next two decades, we conducted research on the river's ecology, including its resident peacock bass. Together with a cadre of students and other science colleagues, we discovered many details about the ecology of this nearly pristine river-floodplain ecosystem. Some of our findings were expected, whereas others were quite surprising. For example, the river is extremely nutrient poor and unproductive, and yet it supports 280 fish species, many of which have impressive abundance. We learned that extensive flooding during the wet season allows fish to exploit terrestrial resources in a vastly expanded aquatic realm. A series of experiments revealed that algae production, the foundation of the aquatic food web, was greater than expected based on our initial impressions of the river's crystal-clear water and clean sandy substrate. We also discovered that bottom-feeding fish called bocachicos (*Semaprochilodus kneri*) compete with microscopic invertebrates buried in the sand to exploit the microscopic bits of algae that accumulate on the sand surface. The abundant bocachicos are migratory and, as prey, provide an important nutritional subsidy for peacock bass and other large predators, thereby helping to support large populations in this nutrient-poor ecosystem (see Chapter 1).

Clear water and clean sandy substrate in the channel of the Cinaruco River in Venezuela's Santos Luzardo National Park. *Photo: K. Winemiller.*

So there we were, pushing and pulling Craig's boat into an isolated lake. We had been assigned "angling duty" for the afternoon. Great work if you can get it! Actually, the objective was to determine the density of peacock bass in a lake that had recently been netted by fishermen illegally. Fishermen from towns along the Orinoco River enter the park, often under cover of darkness, and then use large seine nets to remove fish from shallow lakes in the Cinaruco floodplain. Assessing whether a lake had been exploited in this manner was fairly easy, because the fishermen always left their trash and bycatch (fish of little or no economic value) on the shore. During his 2-year study, Craig recorded how many peacock bass were caught out of 4856 total casts of lures into several lakes, some of which had been recently exploited by netters. In lakes that were never visited by the illegal netters, an average of 6 peacock bass were captured per 100 casts, but in netted lakes the average was less than 2.5 peacock bass per 100 casts, an almost 60% decline in abundance. In a separate analysis of the prey fish communities, Craig found that, when compared to netted lakes, the unnetted lakes had fewer fish within the size interval consumed by adult peacock bass. This was an indication that these voracious predators influence prey fish populations during the dry season [1]. Craig published several reports about aspects of the river's ecology, and he later assumed faculty positions at Florida International University and then North Carolina State University. He eventually shifted his research focus to the ecology of coastal marine systems. One important conclusion from our group's research on the Cinaruco River ecosystem is that peacock bass stocks are extremely vulnerable to overexploitation. Additional evidence showing the effects of excessive harvesting comes from

Craig Layman with royal peacock bass captured during research on the Cinaruco River. *Photo: C. Montaña.*

other regions where peacock bass stocks declined (discussed later, and see Chapter 5), often quite precipitously, soon after the commencement of commercial fishing.

Fisheries—Native stocks

Peacock bass are undeniably South America's most important sport fish. Of course, the Neotropical fish fauna is incredibly diverse, and other large predatory fish targeted by anglers in South America include the dourado (*Salminus brasiliensis*), payara (or peche cachorro, *Hydrolycus* spp.), and large pimelodid catfishes (piraíba or lau lau—*Brachyplatystoma filamentosum*; redtail catfish, pirarara or cajaro—*Phractocephalus hemioliopterus*; jaú—*Zungaro zungaro*; sorubim or bagre rayado—*Pseudoplatystoma* spp.). However, it is the peacock bass that reigns supreme, attracting anglers from across the globe.

A 175-lb lau lau catfish (*Brachyplatystoma filamentosum*) captured from the Siapa River, a clearwater tributary of the Casiquiare in southern Venezuela. *Photo: K. Winemiller.*

The redtail catfish (*Phractocephalus hemioliopterus*) is highly sought by anglers because of its size and strength. *Photo: K. Winemiller.*

Brazil's peacock bass angling tourism is unrivaled, and this is largely attributed to the impressive size and beauty of speckled peacock bass in the upper and middle portions of the vast Rio Negro Basin. Anglers are attracted to this region not only for the chance to catch trophy-size peacock bass, but also for the jungle experience afforded by the region's pristine rivers and rainforests. Sport fishing tourism in the middle Rio Negro began during the early 1990s, and now more than 20 enterprises host anglers, most of them originating from the United States, Japan, Europe, and Brazil. These companies typically practice "catch and release" in hopes of conserving fish stocks [2]. The Amazon region, the state of Amazonas in particular, has the greatest potential for developing tourism based on peacock bass angling. According to the Amazon State Tourism Agency, sport fishing tourism in Brazil attracted 6027 visitors during 2009 and 6630 visitors in 2010. The majority of these anglers came from either São Paulo, Brazil, or the United States and spent between $3500 and $4000 US dollars for sport fishing tour packages. These expenditures provide both direct and indirect economic benefits to people living in the Amazon. Sport fishing operators purchase food and supplies from local sources, and local residents are hired as staff and guides. Anglers and other tourists buy souvenirs, food and travel-related services.

Black water in the Rio Negro Basin sometimes appears red when the sun penetrates shallow water against a white sand background. *Photo: K. Winemiller.*

Most of the angling world records for speckled (*C. temensis*) and Orinoco butterfly peacock bass (*C. orinocensis*) come from tributaries in the upper and middle Rio Negro. To catch a genuine trophy fish, the angler must travel to remote areas where there is no commercial or subsistence fishing and very light pressure from sport fishing. Natural resources within remote regions of the Rio Negro Basin are usually under the control of indigenous communities, and major sport fishing companies negotiate partnerships with these communities. Some of these companies obtain exclusive rights to fish in certain areas, and the communities receive financial benefits by way of employment as crew, guides, and cooks in addition to payments received for fishing rights.

Inside a sporting goods shop in Manaus, Brazil. The speckled peacock bass attracts anglers from all over the world to the Rio Negro. *Photo: K. Winemiller.*

Peacock bass sold as souvenirs to tourists in Manaus. *Photo: K. Winemiller.*

The Ventuari River in Venezuela's Amazonas state supports a significant peacock bass sport fishery. The river's peacock bass (*C. temensis*, *C. orinocensis*, *C. intermedia*) receive fairly moderate pressure from anglers, most of whom practice catch and release. However, about a decade ago, the level of exploitation increased significantly when people began fishing illegally with nets to supply fish to goldminers who also were operating illegally in the region, including within the Yapacana National Park [3]. Except for Guri Reservoir and the upper reaches of the Caroní and Paragua rivers where peacock bass dispersed from Guri, the Guiana Shield rivers of Venezuela's Bolivar state are rarely visited by recreational anglers, and most fishing is for subsistence by local people.

Cerro Autana, a magnificent tepui in the Yapacana National Park in Venezuela's Amazonas state. *Photo: K. Winemiller.*

The Cinaruco River and other Llanos rivers in Venezuela and Colombia appear to have more productive and sustainable peacock bass fisheries compared to the Rio Negro, Upper Orinoco, and other rivers of the Guiana Shield region. For example, the Cinaruco receives more fishing pressure than does the Pasimoni River, located in the remote Rio Negro-Casiquiare watershed in southern Venezuela, and yet the peacock bass stocks in the Cinaruco have been sustained better than those in the Pasimoni. During surveys of peacock bass conducted in the

Pasimoni during 1993, nearly all of the speckled peacock bass were very large, but surveys conducted in 1997, 1998, and 1999 produced much smaller fish (see Chapter 5). Large speckled peacock bass are now uncommon in the Cinaruco as well; however, overall abundance remains high. And yet, productivity even in Llanos rivers has its limits, and peacock bass stocks can be overexploited. In the Aguaro River, two decades of commercial and subsistence fishing decimated the speckled peacock bass population and greatly reduced the Orinoco butterfly peacock bass stock, which came to be dominated by 1-year-old fish [4, 5]. According to anglers who had fished the Aguaro for many years, speckled peacock bass weighing 5–8 kg used to be common. This scenario is the classic response to overexploitation—large fish are eliminated first, and then overall population abundance begins to decline.

Excellent peacock bass fishing can be found in suitable clearwater and blackwater habitats in the vast lowlands of the western Amazon in Colombia, Ecuador, Peru, and Bolivia. The butterfly peacock bass is the only peacock bass species in these regions. *Cichla ocellaris* var. *monoculus* is found throughout the western Amazon, and *C. ocellaris* var. *pleiozona* occurs within the Upper Madeira River in Bolivia and southern Peru. In these regions, peacock bass generally have greater importance for subsistence and commercial fisheries than recreational fishing.

In the coastal region of the Guiana Shield, exceptional fishing for the pond lukunani (*C. ocellaris* var. *ocellaris*) and falls lukunani (*C. cataractae*) can be found in the Essequibo River; however, the sport fishing industry is not very well developed in Guyana. International anglers are now traveling to the Brokopondo Reservoir in Suriname where there is a healthy population of *C. ocellaris*.

Fisheries—Reservoirs

Most of Brazil's many reservoirs constructed for hydropower support peacock bass fisheries. A few of these reservoirs were constructed on rivers within the native ranges of peacock bass (e.g., Balbina, Lajeado, Serra da Mesa, Tucuruí), but the majority of Brazil's existing reservoirs are in southern and eastern Brazil where peacock bass have been introduced (e.g., Cachoeira Dourada, Campo Grande, Corumbá, Itaipu, Juturnaíba, Lajes, Rosana). Although peacock bass generally thrive in reservoirs and often support excellent recreational fishing and sometimes even small-scale commercial fisheries, dams and reservoirs have negatively impacted regional fish diversity and the value of the commercial fish catch. This happens because most of the large, high-value species, such as catfish and dourado, are migratory and require long reaches of free-flowing river to maintain their populations [6, 7]. Venezuela also has peacock bass fisheries in most of its reservoirs, the most notable being the large blackwater reservoir, Guri, and reservoirs constructed on whitewater rivers, such as Camatagua, Boconó-Tucupido, Pao, Masparro, and Las Majaguas.

Peacock bass stocks in reservoirs are vulnerable to overfishing. During the 1980s, butterfly peacock bass weighing nearly 4 kg were commonly caught in Camatagua Reservoir, located in the Llanos region of Venezuela—a short distance from Caracas [8]. Within several years, abundance declined, and fewer large butterfly peacock bass were caught. For example, the largest fish caught during a 1995 fishing tournament weighed only 2.25 kg [9]. This decline in both abundance and fish size was attributed to illegal commercial fishing done at night by people using gillnets and harpoons. Another probable factor was intense fishing pressure by recreational anglers, this owing to the lake's proximity to major urban centers.

An example of wasteful overharvest. These fish were captured in Venezuela during a family holiday outing. Peacock bass stocks can be conserved if only a few fish are harvested, and the remainder released after capture. *Photo: C. Montaña.*

Reservoir productivity varies not only among regions and river basins, but also according to the age of the reservoir. Newly constructed reservoirs typically have high concentrations of dissolved nutrients released from submerged soil and vegetation. A spike in aquatic ecosystem productivity fueled by nutrients released by decomposing vegetation, including timber which has a slower release of nutrients, has been termed the "trophic upsurge." In Venezuela, one of the largest reservoirs for hydroelectric purposes, Guri, is a blackwater, nutrient-poor system that is fed by the Paragua and Caroní rivers. Two peacock bass species (*C. temensis, C. orinocensis*), neither of which is native to the Caroní Basin, were introduced into the reservoir during the final filling stage in 1979. During the 1980s, Guri was renowned for its excellent peacock bass fishing, but soon thereafter stocks declined. Gillnet surveys conducted 1987–89 [10] produced 1.2kg of peacock bass (*C. orinocensis* and *C. temensis* combined) per 100m of net placed within the littoral zone, but surveys conducted during 1993–94 yielded only 0.60kg per 100m [11]. This 50% decline in peacock bass stocks likely resulted from the combined effects of reduction in ecosystem productivity associated with reservoir aging and fishing mortality.

Peacock bass have been widely introduced into reservoirs beyond their native ranges [8, 12–14]. Most peacock bass species thrive in artificial lakes and reservoirs, and frequently dominate the fish communities of these human-constructed ecosystems. These voracious predators generally have a strong influence on populations of other fish species in reservoirs and lakes. In small lakes, peacock bass may virtually eliminate most other fish species, leaving them to resort to cannibalism [15].

Introduced stocks

Peacock bass have been introduced into nonnative waters in South America and other regions of the world. During the early 1960s, *Cichla ocellaris* was introduced into southeastern Florida where other cichlid species were already well established [16, 17]. This management plan was accepted because earlier exotic introductions in the area apparently had not impacted native fishes [18]. The first attempt to establish butterfly peacock bass was unsuccessful, and the fish were later confined to indoor experimental ponds at the Florida Game and Freshwater Fish Commission's Non-native Research Center in Boca Raton. In the 1980s, fingerlings of *C. ocellaris* and *C. temensis* were imported to Florida from several locations in South America and cross bred in an attempt to maximize genetic variability [18]. A stock was eventually established in the canals around Miami, and, based upon the hundreds of photographs posted on the internet, all of these fish appear to be the butterfly peacock bass, *C. ocellaris*. Presently, the butterfly peacock bass is one of the most popular sport fish in southeast Florida [19], with the other being the largemouth bass (*Micropterus salmoides*). Peacock bass received 56% more fishing effort than largemouth bass, with an estimated annual asset value of $6.6 million for the local economy [19]. The Florida peacock fishery is not only popular with anglers, but also with vendors selling t-shirts, tackle and bait, as well as with fishing guides, taxidermists, travel agencies, and outdoor writers. However, if the butterfly peacock

Butterfly peacock bass (*C. ocellaris*) from a south Florida canal. *Photo: John Tinghino.*

bass expands its range in southern Florida, including waters of the Everglades, they could impact native fauna and ecosystems. Peacock bass reportedly have been captured from water bodies in southwestern Florida, but it is unclear if these fish dispersed across the Everglades or were released by aquarists or transported by anglers.

Beginning in 1961, *C. ocellaris* was stocked into reservoirs on the Hawaiian Islands of Kauai and Oahu [20], and subsequently became established in Oahu's Wahiawa Reservoir [21]. In 1966, *C. ocellaris* was introduced to Guam from Hawaii for sport fishing and biological control of other exotic fish. A stock was established, but its ecological effects have not been documented [22]. *Cichla ocellaris* also was stocked into reservoirs in Puerto Rico [23], and fisheries scientists recently proposed to introduce *Cichla temensis* into the island's reservoirs [24]. The rationale for this proposal was that the speckled peacock bass grows much larger than the butterfly peacock bass, and therefore should more effectively control the reservoirs' large tilapia population. Another argument for introducing an additional exotic species into the island's freshwaters was that the native fish fauna is comprised only of marine invaders that have broad distributions, so that the exotic predator poses no existential threat to native biodiversity. In our opinion, *C. temensis* is unlikely to survive in the clear waters that drain from Puerto Rico's mountains, because the water chemistry is unsuitable (e.g., pH 6.7–9.3, with high concentrations of dissolved ions indicated by electrical conductivity from 360 to 12,000 μS/cm) for this blackwater-adapted species. Even the clearwater rivers within *C. temensis*' native distribution tend to be acidic, with pH 5–6 and exceedingly low conductivity from 10 to 50 μS/cm.

Between 1978 and 1984, *Cichla temensis* was introduced into several Texas reservoirs that receive heated water from power plants, but this effort failed. A few of the fish survived and reproduced for a brief period, but by 1992 all of them had died, apparently because they were unable to survive cold winter temperatures [25, 26]. We suspect water quality also could have contributed to the failure of the Texas stockings. As a result of the failure of this experimental introduction, peacock bass were removed from the state's list of prohibited fish species, allowing Texas aquarium hobbyists to obtain these fish legally.

Cichla ocellaris has been introduced into water bodies in Belize, Dominican Republic, France, Kenya, South Africa, Thailand, Vietnam, Malaysia, and Singapore, but apparently the species has only established self-sustaining stocks in Malaysia and Singapore [27]. Many of these introductions were done by anglers or aquarium hobbyists without government approval, and some of these countries lack monitoring programs and laws to manage exotic species [28]. In localities where peacock bass have been introduced outside of their native range, they generally impact the native fish community, often very substantially [12, 29–31]. In Panama's Lake Gatun, the introduction of peacock bass was followed by major changes in aquatic community composition and food web dynamics that affected mosquitos and even human health in communities around the lake [12]. These changes to the Lake Gatun fish community have persisted for 45 years and counting [32].

Planned introductions of peacock bass in Florida, Hawaii, and Puerto Rico have been considered successful, with no serious problems reported [17, 24]. These regions have very few native freshwater fishes, and those that are present tend to be species that are widely distributed in coastal marine, brackish, and fresh waters. Some biologists have claimed that establishment of peacock bass in Florida canals did not affect abundance of native largemouth bass [19, 33]. One study reported that peacock bass can exclude largemouth bass from their normal spawning areas in Florida canals [34]. Peacock bass failed to establish in several Puerto Rican reservoirs, but in other reservoirs on the island they coexist with largemouth bass that were introduced from North America. In some of those reservoirs, peacock bass dominate and displace largemouth bass, but in others they are dominated by largemouth bass [35].

Residents of Miami, Florida, can catch butterfly peacock bass literally in their own backyards and public places. *Photo: John Tinghino.*

When exotic fish are stocked into reservoirs, they inevitably disperse into connected rivers and other water bodies. Peacock bass (principally, *C. ocellaris* var. *kelberi* and *C. piquiti*) have been stocked into most reservoirs and ponds within the Upper Paraná, Paraguay, São Francisco, and other river basins in southern Brazil, and now these species are common in rivers, streams, floodplain lakes, and wetlands throughout this huge region. Research on the impact of these exotic predators on native fish communities has only just begun, but already there is evidence of major effects [30, 36, 37]. For example, when *C. ocellaris* var. *kelberi* was introduced into Rosana Reservoir (Paraná Basin), the number of native fish species declined by 80% and the average density of native fish declined 95% [30].

Since the 1940s, *Cichla orinocensis* and *C. temensis* have been repeatedly transported from locations in the Orinoco Basin and stocked into lakes in nonnative river drainages of the Caribbean Coast, Lake Valencia, and Lake Maracaibo in Venezuela [38]. Escapement from these lakes has led to establishment of self-sustaining populations of *C. orinocensis* in rivers and lakes, but it is unclear if *C. temensis* has become established. The first introductions of *C. orinocensis* in the Maracaibo Basin were along the northeastern shore, and the species eventually spread throughout the giant lake. Since at least 2008, a commercial fishery has been established for the Orinoco butterfly peacock bass in the southern region of the lake near the mouth of the Catatumbo River, this in spite of a national law prohibiting commercial exploitation of peacock bass [38]. We have recently received reports from colleagues in Colombia that *C. orinocensis* and *C. temensis* were introduced and have become established in the Magdalena River Basin. The Magdalena is the major river that runs through the center of Colombia from south to north, emptying into the Caribbean Sea at the city of Barranquilla. The river has extensive floodplains in its middle and lower reaches that should provide good habitat for peacock bass. These invaders likely were transported from rivers of the Orinoco Basin within the Llanos region of Colombia in order to establish sport fisheries closer to major urban centers. Unfortunately, the impacts of these nonnative predators on the endemic fishes of the Magdalena River are impossible to predict at this time.

Care of captive peacock bass

The tropical fish hobby has become increasingly sophisticated in recent decades. Now that we live in the age of internet commerce, it seems almost anything can be purchased and delivered to your door, and this includes live tropical fish. A recent search on the internet identified many fish-import/export companies in several countries with peacock bass species available for purchase and direct delivery to retailers and hobbyists.

How does one keep a peacock bass? Essentially there are two options—pond or aquarium. Unless one lives in a tropical climate or has a means to heat a pond during winter in a temperate climate, a pond probably is not going to be a viable option. Even in a tropical region, maintaining peacock bass in a pond can be a huge challenge because of the necessity for suitable water quality and an adequate food supply. Even in a relatively large pond, a few peacock bass can quickly deplete their food supply, and additional food would need to be provided at regular intervals. I (KW) recently accompanied some friends to fish for peacock bass in some small ponds at a resort near Belém, Brazil. We caught only a few small *C. pinima* (or possibly hybrids between *C. pinima* × *C. ocellaris*) that were small and skinny. The only prey living in the ponds appeared to be a few tilapias, most of which were too large to be consumed by the peacock bass.

Temperature and water quality (pH, conductivity, alkalinity) must be monitored in a peacock bass aquarium. Ammonium, nitrite, and nitrate levels need to be controlled, which requires adequate filtration. Given their huge appetites, peacock bass must be provided a lot of food, and this produces a lot of waste that converts to ammonium, nitrite, and nitrate if not removed.

Peacock bass also require space—lots of it. A 300-gal aquarium is not sufficient, at least not in the long term. Many internet videos posted by hobbyists show multiple peacock bass in tanks no larger than 200 or even 100 gal. We have even seen videos of peacock bass pairs guarding their nests in aquarium that appeared to be no more than 200 gal, so these fish can be amazingly adaptable. Regardless, we recommend at least 500 gal to house two adult peacock bass of any species, and a larger aquarium is even better. This means that very few individuals are capable of properly maintaining peacock bass in their home. Most captive peacock bass will be observed in public aquaria.

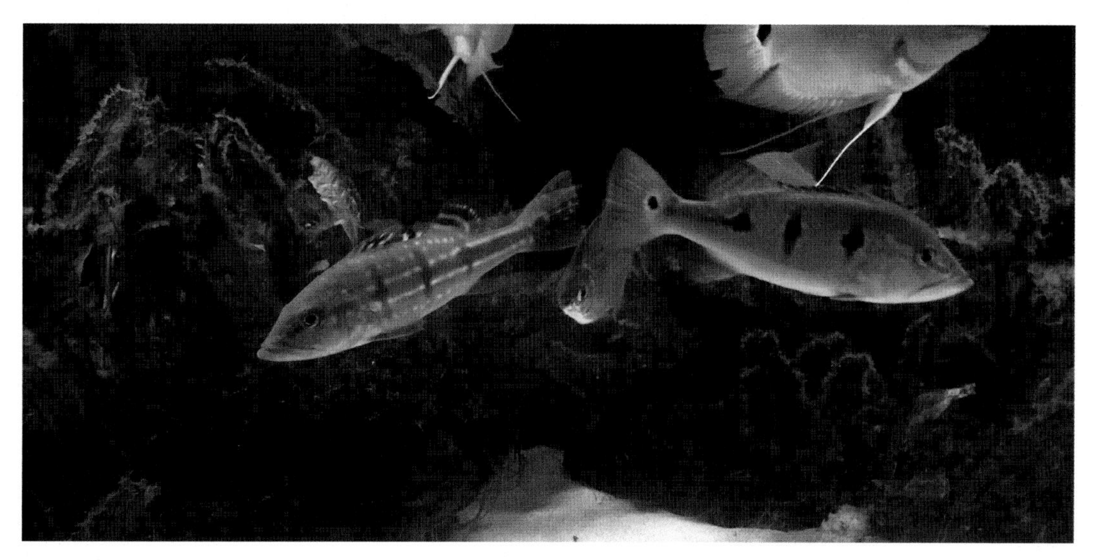

Speckled peacock bass (*C. temensis*, left) and Orinoco butterfly peacock bass (*C. orinocensis*, right) in the authors' home aquarium.

A blue peacock bass (*C. piquiti*, top) and Xingu peacock bass (*C. melaniae*, bottom) in the authors' home aquarium. The fish in the lower left is an eartheater (*Geophagus winemilleri*). *Photo: K. Winemiller.*

Feeding is a challenge, because peacock bass do best on a natural diet—which means fish—preferably live fish. Although peacock bass attack live prey with impressive vigor (be sure the aquarium has a cover or anyone standing near the tank is likely to receive a shower), they can be trained to feed on dead fish or shrimp, and we even have seen videos of peacock bass eating commercial pellets. A big advantage of using dead fish or shrimp is that these foods can be stored frozen and then thawed prior to feeding. Food fish need to be an appropriate size and should lack large spines that could possibly injure the peacock bass. We say this despite the fact that peacock bass in nature sometimes consume fish with spines, such as small catfish and cichlids. The fish should not be fatty or oily, such as smelt, because eventually this makes the peacock bass fat and lethargic, and uneaten pieces of food can foul the water. Depending on the amount of food provided during each feeding, peacock bass do not need to eat every day. However, when hungry after a day or two without food, they often become more aggressive toward their tankmates, especially other peacock bass, which perhaps is a reaction to the presence of potential competitors when food is limiting.

Like all cichlids, peacock bass can display aggression, but usually it is only directed at other peacock bass. In our experience keeping several species of peacock bass in aquaria, it seems that juveniles spend more time chasing and biting other fish. Sometimes this aggression can cause significant damage to fins and scales. Like most cichlids, the aggressor often flares its branchiostegal membranes (folded soft tissue in the throat region) while spreading its dorsal, anal, and caudal fins. Sometimes the dominant fish will approach a subordinate fish and perform an undulating movement while stationary, holding fins erect with branchiostegal membranes flared. This is a threat display observed in virtually all cichlids. The subordinate fish either flees or holds its position while spreading its fins and presenting its flank to the oncoming aggressor. Sometimes this maneuver succeeds in repelling an attack, but also may fail and result in the smaller fish being rammed ("head butted") or bitten. Dominance almost always follows the size hierarchy, with the biggest fish being the bully. If the aquarium is not sufficiently large with enough vegetation, submerged wood or other structures to reduce lines of sight and hinder social interactions, a subordinate fish may be harassed and injured until it assumes a position in one corner of the tank near the surface. There it will remain motionless in hopes of avoiding further interactions with its aggressive tankmate.

In our limited experience keeping seven peacock bass species in aquaria, there appear to be differences in aggression levels among species. Species of *Cichla* Group A (Chapter 11) seem to be the most aggressive toward other peacock bass. Group A includes the *C. temensis* and the other large species. The three smaller species in Group B (*C. ocellaris*, *C. orinocensis*, and *C. intermedia*) definitely show the same behavioral patterns associated with aggression, but they do not seem to be as persistent in their attacks. The Xingu and blue peacock bass (*C. melaniae* and *C. piquiti*) seem to be particularly pugnacious, although not nearly as aggressive and relentless as many of the cichlid species from Central America and Africa. The small, colorful cichlids from the great lakes of the east African Rift Valley are extremely popular in the aquarium hobby, but they tend to be so aggressive that a tankful may induce a feeling of tension rather than relaxation in the observer.

Many photos and videos posted online show aquarium-housed peacock bass with unusual body shapes (very tall and short body, steeply sloping forehead, pointed snout) that are never observed in wild fish. This is caused by an unnatural diet and lack of exercise due to the limited effort required for feeding. Similarly, atypical coloration patterns sometimes are observed in aquarium-raised fish that are rarely seen in the wild, such as dense black spotting, cloudy white or cream spotting all over the head, body, and fins, a black lateral stripe, or the lack of any kind of vertical or horizontal dark pigmentation. Some of these anomalous pigmentation patterns may be due to the kind of diet or water quality conditions during rearing, and some may result from hybridization among captively bred fish.

An unnaturally obese pinima peacock bass that was housed for many years in the authors' home aquarium. *Photo: C. Montaña.*

We (LKW, KOW) have kept seven species of peacock bass in our home aquaria. We buy them as juveniles, because then we know their age and can enjoy them longer. Also, we are able to observe changes in their coloration and behavior as they grow. We start them out in a small tank and feed them mosquitofish from our garden pond. As the fish grow, they are transferred to successively larger aquaria with larger tankmates. Great care must be taken that tankmates are not so large that they might harm the peacock bass, and also not so small that they become their meal. In the penultimate stage, the peacock bass is moved to our 300-gal aquarium where it can live happily with other large fish, at least for a few years. The peacocks continue to grow, and after a while, 300 gal does not seem like an adequate volume to allow fish to display their normal behavioral repertoire. The final step in the progression is to donate the fish to a public aquarium willing to accept a large healthy predator.

A young butterfly peacock bass (*C. ocellaris* var. *kelberi*) in the authors' home aquarium. *Photo: K. Winemiller.*

The same fish from the previous photo several months later is expressing yellow coloration and more black spots on the body and yellow spots on the pectoral, anal, caudal, and rayed dorsal fins. *Photo: K. Winemiller.*

Increasingly, peacock bass are available for purchase from importers in Asia, Europe, Canada, and the United States. Many of these companies sell via the internet, with express delivery of live fish directly to one's door. We have observed seemingly every kind of freshwater tropical fish offered for sale via the internet, and we purchased most of our peacock bass from online sellers. However, we urge caution to any hobbyist contemplating to purchase these amazing fish, because they require a very large aquarium, adequate filtration, and lots of food! It is our belief that fish should not be kept in captivity for long periods unless they can be provided with conditions that simulate the essential features of the species' natural habitat, including water quality, space, habitat structural features, and suitable food and tankmates.

Fisheries conservation

One might assume that fish as voracious, tough, and dominant as peacock bass would be resistant to overfishing; however, this is not the case. Peacock bass are formidable predators and bullies that protect their large broods from hungry neighbors. They also are quite resilient after suffering injury when hooked by anglers or nipped by piranhas. The major limitation for peacock bass growth and population sustainability is the low productivity of the blackwater and clearwater rivers and lakes where they live. These tropical ecosystems tend to have very low concentrations of dissolved nutrients that support growth of algae and aquatic vegetation. These plants are the foundation of aquatic food webs, and what little production exists in the aquatic environment has to trickle upward through the aquatic food web to support peacock bass biomass. As we discussed in the preceding chapter, it takes at least 5 years to grow an impressive butterfly or royal peacock bass, and it takes about 10 years to grow a trophy speckled peacock bass. This is important to appreciate, because when the largest fish are harvested (or hooked so many times that they may die from injuries), it takes a long time to replace those older fish.

The key to managing a sustainable peacock fishery is limited fishing pressure and harvest. This may seem intuitive and should apply to any fish stock, but in unproductive tropical ecosystems, fishing pressure has to be extremely light, and even a small amount of sustained harvest will lead to population decline. The sport fishing tourism companies operating in the Rio Negro Basin figured this out a long time ago, probably by trial and error. It did not take long for them to realize that when anglers return to the same fishing location time and time again, the catch declines rapidly. These companies cannot afford to send clients home dissatisfied, and they soon realized that to keep anglers happy, the solution was to alternate their trips among multiple fishing locations. This requires advanced planning and extra expense, but it is crucial for sustained peacock bass fishing success. This strategy avoids the scenario we described for the Pasimoni River in Venezuela (Chapter 5) whereby international anglers repeatedly fished the same locations and crashed the trophy fishery in less than 5 years. Another consequence of repetitive visits to the same location is that this attracts the attention of local people who begin to ponder why so many foreigners are suddenly invading their area. Sooner or later they conclude that something in that river must be extremely valuable. At first the locals thought the "gringos" were searching for gold in the Pasimoni River, but when they realized visitors were pursuing big fish, they decided they had better get some also, whether or not they really needed them. Although accessing the sites required either gasoline (expensive and hard to obtain) or several days of paddling, the inconvenience must be worth it if the fishing was so great. This illustrates why it is important for the sport fishing companies to engage local communities in business agreements that allow everybody to benefit from the enterprise.

Lake in the floodplain of the Pasimoni River, a blackwater tributary of the Casiquiare in southern Venezuela. This pristine lake produced many huge speckled peacock bass in the early 1990s, but by the late 1990s, the average size of fish had declined significantly from fishing. *Photo: K. Winemiller.*

The management model of shifting fishing pressure among locations seems to work well in the Rio Negro Basin, and that is because there are very few people living in this nutrient-poor wilderness. Fishing locations are plentiful, and relative to the immense size of the basin, there are few outsiders entering to fish them out. When there are cooperative agreements with outfitters, the local people maintain vigilance over the prime fishing areas, because they receive financial benefits from sport fishing. Key to the success of this kind of community-based fishery management is the fact that peacock bass don't migrate, at least not very far (see Chapter 5).

What about commercial and subsistence fisheries in the Amazon? Markets in Belém, Santarem, Manaus, Tefé, Iquitos, and other Amazon cities are well stocked with peacock bass, including *C. ocellaris*, *C. temensis*, and *C. pinima*. There must be a lot of fish somewhere. For now, the total human population in the Amazon is relatively small given the vast area encompassed by the Amazon Basin, but this human population is growing very rapidly. Recently, Brazil's government has advanced policies aimed at promoting economic growth in the Amazon, and as a result, the region's population has grown from 1.4 million to 15.9 million inhabitants in less than a century. For now, there are reasonably healthy peacock bass stocks in regions of the Amazon that are far from large towns and cities. Peacock bass of truly impressive size are rarely seen in markets, except perhaps in villages in very remote regions. Peacock bass stocks usually are depleted in areas close to large towns and cities in the Amazon, and fish in their markets sometimes are transported from remote areas.

Peacock bass are delicious no matter how they are cooked. These butterfly peacock bass were roasted over a campfire and provided a feast for researchers during an expedition in Guyana. *Photo: D. Bloom.*

Speckled peacock bass (*C. temensis*) for sale in the Manaus fish market. *Photo: L. Kelso-Winemiller.*

Commercial fishing for peacock bass is allowed in Brazil and several other Amazonian countries but is banned in Venezuela and Guyana (subsistence fishing is permitted by indigenous communities). Enforcement of the fishing laws is severely lacking in most South American countries, and illegal commercial fishing for peacock bass has existed for decades in Venezuela. That country's Ley de Pesca (Fishing Law) established regulations for peacock bass recreational, artisanal, and subsistence fisheries, and set a minimum size of 30 cm for all *Cichla* species while limiting daily possession to five fish per person. Brazil's Instituto Brasileiro do Meio Ambiente (Environmental Institute) established a minimum capture size of 35 cm for all *Cichla* species and possession of no more than 5 kg per person per day in waters where peacock bass are native. In rivers where nonnative peacock bass have been introduced, the law allows fishing, capturing, and transporting any size and amount of peacock bass.

Fresh from the Rio Negro; butterfly peacock bass (*C. ocellaris* var. *monoculus*) for sale dockside in Manaus. *Photo: L. Kelso-Winemiller.*

Local fisherman with his commercial catch from the lower Branco River in Brazil. These fish will be sold to a fishmonger who transports them to the market in Manaus. *Photo: K. Winemiller.*

Indigenous fisherman from the Ventuari River, Venezuela, with his catch of speckled peacock. This is subsistence fishing for local consumption. *Photo: C. Montaña.*

Peacock bass stocks have been severely depleted in most rivers of the Venezuelan Llanos [1, 39, 40], and there seems to be little prospect for recovery of these stocks given the dire economic situation in that country at present. At last report, the Cinaruco River still had a viable sport fishery for peacock bass despite illegal commercial fishing; however, large trophy fish are rare. Commercial fishermen often capture peacock bass from reservoirs using gill nets (stationary nets that trap fish as they attempt to swim through their mesh).

Stuart Willis with speckled peacock bass sold illegally in the market of a town in the Venezuelan Llanos. *Photo: C. Montaña.*

In reservoirs, the sustainability of peacock bass stocks depends on the nutrient status and productivity of the watershed, the age of the reservoir, and fishing pressure. Reservoirs built on blackwater rivers, such as Guri in Venezuela, are extremely nutrient poor and unproductive. Stocks in these reservoirs are highly sensitive to harvest. In more productive reservoirs located in nutrient-rich watersheds, peacock bass populations can withstand greater levels of harvest; however, fish will tend to be small unless catch and release is adopted. Some reservoirs may be sufficiently productive to sustain subsistence fishing or even a low level of commercial harvest. This seems to be the case for Lajeado Reservoir on the Tocantins River.

Puerto Rico allows anglers to possess 8 peacock bass per day. Florida's possession limit for peacock bass is 2 per angler per day, with only one fish greater than 17 in., and catch and release is encouraged. Despite our tremendous appreciation for these amazing fish, we do not support the introduction of peacock bass into nonnative waters. Despite the short-term economic benefits, there is significant risk of impacts to native biodiversity and ecosystem services that benefit people. Once established, exotic species can more easily be transported to other locations, either intentionally or unintentionally, thereby increasing the risk of large-scale invasion and impacts. For us, much of the allure of fishing for peacock bass is the experience of traveling to exotic locations where one finds an unspoiled tropical environment and interesting culture. We feel the same way about North American fish that have been stocked in nonnative regions of the United States where they have established and spread. Exotic fish introduced from the Mississippi River Basin have severely impacted endemic native fishes in rivers of the western United States, including the Colorado, Colombia, and Sacramento-San Joaquin rivers.

We know some accomplished anglers who travel all over the world to catch legendary fish, and some of them routinely travel to Thailand to fish lakes stocked with South American game fish, including peacock bass, tambaqui (*Piaractus brachypomus*), redtail catfish (*Phractocephalus hemioliopterus*), and even giant pirarucú (*Arapaima gigas*). We have conducted research in Southeast Asia and find that region's native fishes to be as fascinating as they are diverse. Some of the world's largest freshwater fishes (e.g., the giant Mekong catfish, *Pangasianodon gigas*) and fiercest predators (e.g., the voracious and rather frightening giant snakehead, *Channa micropeltes*) are native to Southeast Asia. Stocking lakes with exotic predators not only poses a great risk to local biodiversity and ecosystems, but also it is unappealing to us as naturalists. To catch South American fishes, one should travel to South America to experience all of its wonders. To see Asian fish, visit Asia and enjoy the region's natural history and culture. In the freshwater canals of south Florida, one should find native largemouth bass (*Micropterus salmoides*) and Florida gar (*Lepisosteus platyrhincus*), impressive fish in their own right.

Sport fisheries management and anglers are not the only ones to blame for introductions of exotic fish species. The aquarium fish industry and hobbyists are responsible for many introductions, both intentional and unintentional. For example, a single *Cichla ocellaris* was reported caught by an angler in Arizona, and this fish likely was released by a hobbyist. The canals of south Florida are full of exotic fishes, most of which were released by well-meaning hobbyists. Several years ago, one of us (KW) accompanied a group of students to fish in the Big Cypress National Preserve to obtain fish for laboratory research. We caught several oscars (*Astronotus ocellatus*), tilapia (*Tilapia mariae*), and various other tropical fish. Oscars are fascinating fish and strong fighters on the end of a line, but it seemed more than a little strange to catch them in North American waters.

An oscar (*Astronotus ocellatus*) caught from a floodplain lake in the Amazon. This species is popular in the aquarium hobby and was introduced into freshwater habitats in south Florida where it is now common. *Photo: K. Winemiller.*

Epilogue

As we explained at the beginning of this book, and hopefully have demonstrated throughout its pages, peacock bass are special. Not only are they among of the world's premier sport fish, but also they are stunningly beautiful with their brilliant colors, vivid eyespots, and bold bars, blotches, spots, and stripes. Their streamlined muscular bodies are as elegant as they are powerful. These spectacular fish almost seem better suited for life on coral reefs instead of dark jungle waters. Peacock bass are a bit like chameleons, with their color pattern changing according to their environment, life stage, reproductive state, and mood. The diligence with which peacock bass defend their offspring is rivaled only by the voracity with which they attack their hapless prey. Perhaps most importantly, peacock bass are critical components of aquatic food webs in their native rivers of the Amazon, Orinoco, and Guiana Shield regions of South America. We should appreciate their beauty, behavior, sporting qualities, and role in healthy tropical ecosystems, and work to conserve them for the health of the planet and enjoyment of future generations.

A beautiful fire peacock bass (*C. mirianae*) being released unharmed back into the clear waters of the Sáo Benedito River, Brazil. *Photo: K. Winemiller.*

References

[1] C.A. Layman, K.O. Winemiller, Patterns of habitat segregation among large fishes in a Neotropical floodplain river, Neotrop. Ichthyol. 3 (2005) 111–117.

[2] C.E.C. Freitas, A.A.F. Rivas, A pesca e os recursos pesqueiros na Amazônia Ocidental, Ciênc. Cult. 58 (3) (2006) 30–32.

[3] Y. Lesenfants, M. Molinillo, Manaka: Paraiso de pesca y puerta de entrada a la region natural Duida-Ventuari, Rev. Divulg. Cient. Fund. Cisneros 1 (2001) 8–28.

[4] D.B. Jepsen, K.O. Winemiller, D.C. Taphorn, D. Rodriguez Olarte, Age structure and growth of peacock cichlids from rivers and reservoirs of Venezuela, J. Fish Biol. 55 (1999) 433–450.

[5] D. Rodriguez-Olarte, D.C. Taphorn, Distribucion y abundancia de pavones (*Cichla orinocensis* y *C. temensis*, Pisces:Cichlidae) en un humedal de los Llanos Centrales en Venezuela, Vida Silv. Neotrop. 8 (1–2) (2001) 43–50.

[6] A.C. Ribeiro, Tectonic history and the biogeography of the freshwater fishes from the coastal drainages of eastern Brazil: an example of faunal evolution associated with a divergent continental margin, Neotrop. Ichthyol. 4 (2006) 225–246.

[7] D.J. Hoeinghaus, A.A. Agostinho, L.C. Gomes, E.K. Okada, F.M. Pelicice, E.A.L. Kashiwaqui, J.D. Latini, K.O. Winemiller, Effects of river impoundment on ecosystem services of large tropical rivers: embodied energy and market value of artisanal fisheries, Conserv. Biol. 23 (2009) 1222–1231.

[8] R. Ochoa, Los pavones en Camatagua, Natura 96 (1993) 40–41.

[9] R. Ochoa, La pesca de pavones en Camatagua, Rev. Divulg. Cientif. Fund. Cisneros 1 (2001) 70–77.

[10] D.F. Novoa, J. Koonce, F. Ramos, La ictiofauna del lago Guri: composición, abundancia y potencial pesquero. II. Evaluación del potencial pesquero del lago Guri y estrategias de ordenamiento pesquero, Mem. Soc. Cienc. Nat. La Salle 49 (1989) 159–197.

[11] J.D. Williams, K.O. Winemiller, D.C. Taphorn, L. Balbas, Ecology and status of piscivores in Guri, an oligotrophic tropical reservoir, N. Am. J. Fish Manag. 18 (1998) 274–285.
[12] T.M. Zaret, R.T. Paine, Species introduction in a tropical lake, Science 182 (1973) 449–455.
[13] D. Novoa, Análisis histórico de las pesquerías del Río Orinoco, in: D. Novoa (Ed.), Los recursos Pesqueros del Río Orinoco y Su Explotación, CVG, Caracas, Venezuela, 1982.
[14] K.O. Winemiller, Ecology of peacock cichlids (*Cichla* spp.) in Venezuela, J. Aquaric. Aquat. Sci. 9 (2001) 93–112.
[15] L.N. Santos, A.F. Gonzáles, F.G. Araújo, Dieta do tucunaré-amarelo *Cichla monoculus* (Bloch & Schneider) (Osteichthyes, Cichlidae), no reservatório de Lajes, Rio de Janeiro, Brasil, Rev. Bras. Zool. 18 (2001) 191–204.
[16] W.R. Courtenay Jr., C.R. Robins, Fish introductions: good management, mismanagment, or no management? CRC Crit. Rev. Aquat. Sci. 1 (1) (1989) 159–172.
[17] P.L. Shafland, Exotic fishes of Florida, Rev. Fish. Sci. 4 (2) (1994) 101–122.
[18] P.L. Shafland, Introduction and establishment of a successful butterfly peacock fishery in southeast Florida canals, Am. Fish. Soc. Symp. 15 (1995) 443–451.
[19] P.L. Shafland, The introduced butterfly peacock (*Cichla ocellaris*) in Florida. II. Food and reproductive biology, Rev. Fish. Sci. 7 (1999) 95–113.
[20] J.A. Maciolek, Exotic fishes in Hawaii and other islands of Oceania, in: W.R. Courtenay Jr., J.R. Stauffer Jr. (Eds.), Distribution, Biology, and Management of Exotic Fishes, Johns Hopkins University Press, Baltimore, MD, 1984, pp. 131–161.
[21] W.S. Devick, Life history study of the tucunaré *Cichla ocellaris*, in: Federal Aid in Sportfish Restoration Project F-9-1, Job Completion Report, Hawaii Department of Land and Natural Resources, Honolulu, 1972.
[22] R.L. Welcomme, International Introductions of Inland Aquatic Species, Food and Agriculture Organization of the United Nations, Rome, 1988. Fisheries Technical Paper 294.
[23] D. Erdman, Exotic fishes in Puerto Rico, in: W.R. Courtenay Jr., J.R. Stauffer Jr. (Eds.), Distribution, Biology, and Management of Exotic Fishes, Johns Hopkins University Press, Baltimore, MD, 1984.
[24] J.W. Neal, J.M. Bies, C.N. Fox, C.G. Lilyestrom, Evaluation of proposed speckled peacock bass *Cichla temensis* introduction to Puerto Rico, N. Am. J. Fish Manag. 37 (5) (2017) 1093–1106.
[25] G.P. Garrett, Status report on peacock bass (*Cichla* sp.) in Texas, in: Annual Proceedings of the Texas Chapter, American Fisheries Society, 1982.
[26] R.G. Howells, G.P. Garrett, Status of some exotic sport fishes in Texas waters, Tex. J. Sci. 44 (3) (1992) 317–324.
[27] J.H. Liew, H.T. Heok, D.C.J. Yeo, Some cichlid fishes recorded in Singapore, Nat. Singap. 5 (2012) 229–236.
[28] B.C. Tan, K.S. Tan, Singapore, in: N. Pallewatta, J.K. Reaser, A.T. Gutierrez (Eds.), Invasive Alien Species in South-Southeast Asia: National Reports & Directory of Resources, Global Invasive Species Programme, Cape Town, South Africa, 2003.
[29] A.O. Latini, M. Petrere Jr., Reduction of a native fish fauna by alien species: an example from Brazilian freshwater tropical lakes, Fish. Manag. Ecol. 11 (2004) 71–79.
[30] F.M. Pelicice, A.A. Agostinho, Fish fauna destruction after the introduction of a non-native predator (*Cichla kelberi*) in a neotropical reservoir, Biol. Invasions 11 (2009) 1789–1801.
[31] R.F. Menezes, J.L. Attayde, G. Lacerot, S. Kosten, L. Coimbra de Souza, L.S. Costa, E.H. Van Nes, E. Jeppesen, Lower biodiversity of native fish but only marginally altered plankton biomass in tropical lakes hosting introduced piscivorous *Cichla cf. ocellaris*, Biol. Invasions 14 (2012) 1353–1363.
[32] D.M.T. Sharpe, L.F. De Leon, R. Gonzalez, M.E. Torchin, Tropical fish community does not recover 45 years after predator introduction, Ecology 98 (2017) 412–424.
[33] J.E. Hill, Effects of Introduced Peacock Cichlids *Cichla ocellaris* on Native Largemouth Bass *Micropterus salmoides* in Southeast Florida (Ph.D. dissertation), University of Florida, Gainesville, 2003.
[34] L. Nico, P. Fuller, Spatial and temporal patterns of nonindigenous fish introductions in the United States, Fisheries 24 (1999) 16–27.
[35] J.W. Neal, R.L. Noble, C.G. Lilyestrom, Evaluation of the ecological compatibility of butterfly peacock cichlids and largemouth bass in Puerto Rico reservoirs, Trans. Am. Fish. Soc. 135 (2006) 288–296.
[36] A.A. Agostinho, F.M. Pelicice, L.C. Gomes, Dams and the fish fauna of the Neotropical region: impacts and management related to diversity and fisheries, Braz. J. Biol. 68 (4) (2008) 1119–1132.
[37] K.E. Kovalenko, E.D. Dibble, A.A. Agostinho, G. Cantanhêde, R. Fugi, Direct and indirect effects of an introduced piscivore, *Cichla kelberi* and their modification by aquatic plants, Hydrobiologia 638 (2010) 245–253.

[38] O.M. Lasso-Alcalá, G. Andrade de Pasquier, C. Hoyos, B. Bottini, M. Hernández Nácar, Sobre la introducción de los pavones, *Cichla orinocensis* y *C. temensis* (Perciformes, Cichlidae), en la Cuenca del Lago de Maracaibo, Venezuela, Anartia 26 (2016) 31–50.
[39] D. Rodríguez-Olarte, D.C. Taphorn, Ecologia trofica de *Cichla orinocensis* Humbolt 1833 (Pisces, Teleostei, CIchlidae) en un humedal de los llanos centrales de Venezuela, Biollania 13 (1997) 139–163.
[40] D.J. Hoeinghaus, C.A. Layman, D.A. Arrington, K.O. Winemiller, Movement of *Cichla* species (Cichlidae) in a Venezuelan floodplain river, Neotrop. Ichthyol. 1 (2003) 121–126.

Index

Note: Page numbers followed by *f* indicate figures and *t* indicate tables.

A

Academy of Natural Sciences of Philadelphia, 177–178
Acestrorhynchus falcirostris (Pike characins), 107
Aguaro River, 12*f*
Amazon River, 13*f*, 246–249, 247–249*f*
Angling, 271–274
Apex predator, 10
Apistogramma species (Dwarf cichlids), 7*f*
Apure River, 13*f*
Aquarium hobby, 280, 284, 296*f*
Aquatic food web, 269–270
Araguaia cichlid (*Cichlasoma araguaiense*), 165
Araguaia-Tocantins Basin, 156, 259–260, 260*f*
Asian snakehead, 138
Astronotus ocellatus (oscar), 66–67, 296*f*

B

Background coloration, 24–25, 24–26*f*
Banded peacock bass, 110*f*
Behavior, 10, 23–24
Belo Monte Dam, 174, 176*f*, 187, 189–190
Black caiman (*Melanosuchus niger*), 204*f*
Blackwater river, 107, 108*f*, 110, 114*f*, 118, 126
Blue peacock bass. *See Cichla piquiti* (Blue peacock bass)
Bocachicos (*Semaprochilodus kneri*), 269–270
Brachyplatystoma filamentosum (Catfish), 272*f*
Branco River, 36, 40*f*, 46*f*, 50–51, 292*f*
Brazil, 177–178
  Branco River, 72–73
  Lajeado Reservoir, 155*f*
  natural distribution, *Cichla orinocensis*, 72*f*
  Palmas city, 154–155
  Rio Negro, 84
  Upper Paraná River near Porto Rico, 160*f*
Brazilian Shield, 175–176
Brood guarding, 35*f*, 54, 58–59, 61
Butterfly peacock bass. *See Cichla ocellaris* (butterfly peacock bass)

C

Cachoeira Dourada Reservoir in Upper Paraná Basin, 166
Caiman crocodilus (Spectacled caiman), 82, 82*f*
Cannibalism
  alternative prey availability, 54
  Cichla orinocensis (Orinoco butterfly peacock bass), 81
  Cichla temensis (speckled peacock bass), 124–125, 124–125*f*
  gamete production and brood guarding, 54
Captive peacock bass, 282–287, 283*f*, 285–286*f*
Casiquiare River, 34–35, 46, 46*f*
Catfish (*Brachyplatystoma filamentosum*), 272*f*
Caura River
  Guiana Shield, southern Venezuela, 92
  rocky shoals, 103*f*
  royal peacock bass, 94*f*
  Venezuela's Bolivar State, 92
Cichlidae, 8–9, 28
Cichla cataractae (Falls lukunani)
  Amazon and Orinoco basins, 217
  distribution, 222–225, 223–224*f*
  Essequibo River, 218*f*
  feeding, 225, 225*f*
  growth, 225–226, 226*f*
  identification, 218–222, 219–222*f*
  reproduction, 227–229, 228–229*f*
  Rewa River, 216
  rocky shoals, 216, 217*f*
Cichla intermedia (royal peacock bass)
  Cinaruco River, 244*f*
  distribution, 97–99, 97–99*f*
  growth, 100–101, 100*f*
  identification, 94–97, 95–96*f*
  with large nuchal hump, 93–94*f*
  Manipitare River, 240*f*
  river channel habitats, 93
  Venezuela's Caura River, 92*f*
Cichla melaniae (Xingu peacock bass)
  distribution, 180–184, 181–184*f*
  growth, 185–187, 186*f*
  hydropower, impacts of, 189–192, 190–191*f*
  identification, 178–180, 178–180*f*
  reproduction, 188–189, 189*f*
Cichla mirianae (Fire peacock bass)
  distribution, 201–205, 202–205*f*
  feeding, 205–210, 206–209*f*
  growth, 210
  identification, 199–201, 199–201*f*
  reproduction, 212–214, 212–214*f*
  São Benedito River, 197, 198*f*, 256*f*, 297*f*

Cichla ocellaris (butterfly peacock bass)
Araguaia-Tocantins Basin, 41
breeding pair of, 40–41*f*
characteristics, 35*f*
color patterns, 37
distribution, 72–78, 72–77*f*, 79*f*
feeding, 78–82, 79–80*f*
fisheries biologists and anglers, 34
floodwaters, 36*f*
growth of, 56–57, 82–84, 83*f*
identification, 39–46, 68–72, 68–71*f*
immature, Pirara River, 37*f*
juveniles, 43, 44*f*
origin of, 39
reproduction, 84–88, 85–88*f*
Rupununi River, 35*f*, 36
south Florida canal, 279*f*
species description, 1801, 39*f*
"species" *vs.* "varieties or subspecies,", 37
subadult and adult, 40
varieties, 39
Cichla ocellaris var. *monoculus*, 43*f*, 45*f*
Cichla ocellaris var. *kelberi* (young butterfly peacock bass), 42*f*, 45*f*, 286*f*
Cichla pinima (Pinima peacock bass)
distribution, 143–149, 143–145*f*, 147–148*f*
feeding, 149, 149*f*
growth, 149–150, 149*f*
identification, 139–143, 140–142*f*
Jari River, 233*f*
reproduction, 150–152, 151–152*f*
Cichla piquiti (Blue peacock bass)
Araguaia River, 156
Araguaia-Tocantins Basin, 260*f*
"cerrado" region, 154–155
distinct color, 157*f*
distribution, 161–165, 161–163*f*
growth, 166–167, 166–167*f*
home aquarium, 283*f*
identification, 157–161, 157–160*f*
Lajeado Reservoir in Brazil, 155*f*
Palmas, Brazilian state of Tocantins, 154–155
reproduction, 169–170, 169–170*f*
Cichlasoma araguaiense (Araguaia cichlid), 165
Cichla temensis (speckled peacock bass), 1–4, 3*f*, 283*f*, 290*f*
black waters of Rio Negro, 253*f*
clearwater and blackwater rivers, 246*f*
distribution, 115–121, 115–120*f*
feeding, 121–125, 121–123*f*
growth, 125–126, 126*f*
identification, 111–115, 111–114*f*
reproduction, 130–134, 131–133*f*
Cichlids
class Actinopterygii and order Cichliformes, 232
Crenicichla lenticulata, 233*f*
neotropical, 232, 232*f*
Retroclus xinguensis, 233*f*
in South America, 232–233
Cinaruco-Capanaparo National Park, 269–270
Cinaruco River, 95–96*f*, 99, 100*f*, 101, 270–271*f*, 271–272
clearwater tributary, 1–4, 3*f*
fine speckled peacock bass, 1–4, 3*f*
Llanos region of Venezuela, 1–4, 2*f*
Clades, 234–235, 236*f*
Clearwater, Lower Xingu River, 174
Cojedes River, 66–67
Coloration patterns, 111, 126*f*, 130–131, 138
Commercial fishery, 281–282
Competitors, 15–21, 15–23*f*
Crenicichla lenticulata (pike cichlid), 7*f*, 66–67, 233*f*

D

Distribution
Cichla cataractae (Falls lukunani), 222–225, 223–224*f*
Cichla intermedia (royal peacock bass), 97–99, 97–99*f*
Cichla melaniae (Xingu peacock bass), 180–184, 181–184*f*
Cichla mirianae (Fire peacock bass), 201–205, 202–205*f*
Cichla ocellaris (butterfly peacock bass), 46–52
Cichla orinocensis (Orinoco butterfly peacock bass), 72–78, 72–77*f*, 79*f*
Cichla pinima (Pinima peacock bass), 143–149, 143–145*f*, 147–148*f*
Cichla piquiti (Blue peacock bass), 161–165, 161–163*f*
Cichla temensis (speckled peacock bass), 115–121, 115–120*f*
Dwarf cichlids (*Apistogramma* species), 7*f*

E

Eartheaters (*Geophagus* and *Satanoperca* species), 7*f*, 165
Essequibo River Basin, 216, 218*f*, 221*f*, 222, 224–227*f*
Evolutionary lineages, 145*f*
anatomical/coloration characteristics, 235–237
Cichla melaniae, Group A lineage, 237*f*
Cichla species, DNA sequence data, 235, 235*f*
C. ocellaris var. *monoculus,* Casiquiare River, 239*f*
different colors, *Cichla pinima*, 238*f*
distribution map, 236*f*
genetic divergence, 235–237
Jari River, *Cichla pinima*, 238*f*
nucleotide sequencing and statistical analysis, 234
regional stocks, 239, 239*f*
Royal peacock bass (*C. intermedia*), 240*f*

F

Falls lukunani. *See Cichla cataractae* (Falls lukunani)
Feeding, 121–125, 121–123*f*
Cichla cataractae (Falls lukunani), 225, 225*f*

Cichla intermedia (royal peacock bass), 16f, 99–100
Cichla melaniae (Xingu peacock bass), 184–185, 185f
Cichla mirianae (Fire peacock bass), 205–210, 206–209f
Cichla ocellaris (butterfly peacock bass), 53–56
Cichla orinocensis (Orinoco butterfly peacock bass), 78–82, 79–80f
Cichla pinima (Pinima peacock bass), 149, 151f
Cichla piquiti (Blue peacock bass), 165–166, 165f
Fire peacock bass1. See Cichla mirianae (Fire peacock bass)
Fisheries
conservation, 287–296, 288–294f, 296f
native stocks, 272–277, 273–276f
reservoirs, 277–278, 278f
Fish stocking, 55f
Fin coloration, 30, 31f
Flag cichlids (Mesonauta species), 7f
Floodplains, 1–4, 7f, 10–11, 19f
Food webs, 54
Freshwater dolphins (Inia geoffrensis), 82

G
Genetic variation, 130
Geographic variation, Cichla ocellaris (butterfly peacock bass), 46
Geophagus and Satanoperca species (Eartheaters), 7f, 165
Giant river otters (Pteronura brasiliensis), 82
Guiana shield coastal rivers, 92, 97–98, 107, 260–263, 261–263f
Guri Reservoir, 67, 70, 74, 81, 84
Guyana, 216, 218f, 222, 223f

H
Habitat
Cichla cataractae (Falls lukunani), 222–225, 223–224f
Cichla intermedia (royal peacock bass), 97–99, 97–99f
Cichla melaniae (Xingu peacock bass), 180–184, 181–184f
Cichla mirianae (Fire peacock bass), 201–205, 202–205f
Cichla ocellaris (butterfly peacock bass), 46–52
Cichla orinocensis (Orinoco butterfly peacock bass), 72–78, 72–77f, 79f
Cichla pinima (Pinima peacock bass), 143–149, 143–145f, 147–148f
Cichla piquiti (Blue peacock bass), 161–165, 161–163f
Cichla temensis (speckled peacock bass), 115–121, 115–120f
Headstanders (Anostomidae), 7f
Heros severus (Severums), 66–67, 138
Hoatzin/stinkbird (Opisthocomus hoazin), 196f
Hoplarchus psittacus (Parrot cichlid), 66–67, 138
Hybridization, 263–266, 264–266f
Hypancistrus zebra (zebra pleco), 177f

I
Inia geoffrensis (Freshwater dolphins), 1
The International Gamefish Association (IGFA), 56
Introduced stocks, 279–282, 279f, 281f

J
Jacundá River (Amazon), 38f
Jari River (Amazon), 38f
Juruá River (Amazon), 38f

L
Lajeado Reservoir (Tocantins), 38f
Large catfishes (Pimelodidae), 7f
Las Majaguas Reservoir, 67, 67f, 81, 84
Lateral band (stripe), 27, 28f
Lateral blotches and ocellated markings, 28–29, 29f
Life history, Cichla orinocensis (Orinoco butterfly peacock bass), 68
Llanos region, 84–85
Lower Amazon, 143, 145–146
Lower Branco River (Amazon), 38f
Lower Xingu River (Amazon), 38f

M
Madeira River, 250–251, 250f
Melanosuchus niger (Black caiman), 204f
Metynnis spp. (Silver dollars), 165
Morphological traits, 23–24

N
Natural history, peacock bass, 4–11
Nesting, 35f, 40f, 56f, 59f, 61

O
Opisthocomus hoazin (hoatzin/stinkbird), 196f
Orinoco butterfly peacock bass. See Cichla orinocensis (Orinoco butterfly peacock bass)
Orinoco Basin, 72–75, 243–246, 243–246f
Oscar (Astronotus ocellatus), 66–67, 296f

P
Pamoni River (Casiquiare, Amazon-Orinoco), 38f
Parasites, 15–21, 15–23f
Parrot cichlid (Hoplarchus psittacus), 66–67, 138
The Pasimoni River, 108–110f, 110
"Pavón cinchados,", 1–4
Peacock bass species
background coloration, 24–25, 24–26f
care and breeding of, 7–8
Guiana and Brazilian shields, 5f
ichthyologists or aquatic ecologists, 4–5
morphological features, 24f
new species and disagreement among ichthyologists, 23–24

Peacock bass species *(Continued)*
Pasiba River, floodplain of, 7*f*
rivers and lakes, 4–5
small ornamental fishes, 7*f*
South America, river basins of, 6*f*
vertical dark bars, 26–27, 27*f*
Phractocephalus hemioliopterus (Redtail catfish), 273*f*
Phylogeny
of *Cichla*, 236*f*
evolutionary units, 234–235
and population genetics, 234
Pike characins (*Acestrorhynchus falcirostris*), 107
Pike cichlids (*Crenicichla lugubris*), 7*f*, 66–67
Pinima peacock bass. *See Cichla pinima* (Pinima peacock bass)
Piranhas (*Serrasalmus* species), 7*f*, 107
Plecos (family Loricariidae), 7*f*
Population structure and abundance
Cichla intermedia (royal peacock bass), 101–102, 101*f*
Cichla ocellaris (butterfly peacock bass), 52–53
Population abundance and structure, 84
Cichla cataractae (Falls lukunani), 226–227, 227*f*
Cichla melaniae (Xingu peacock bass), 187–188, 187–188*f*
Cichla mirianae (Fire peacock bass), 210–212, 210–211*f*
Cichla pinima (Pinima peacock bass), 150, 150*f*
Cichla piquiti (Blue peacock bass), 167–168, 168*f*
Cichla temensis (speckled peacock bass), 126–130, 127–130*f*
Postorbital stripe and blotches, 29, 30*f*
Predation mortality, 82, 124–125, 124–125*f*
Predators, 15–21, 15–23*f*
Pseudoplatystoma tigrinum (Tiger shovelnose catfish), 138
Pteronura brasiliensis (Giant river otters), 82

R

Rapids
and cascades, 175–176
of Volta Grande region, 177*f*, 180–181
of the Xingu River, 175*f*, 182*f*, 186*f*
Red pike cichlid, 8*f*
Redtail catfish (*Phractocephalus hemioliopterus*), 273*f*
Reproduction
Cichla cataractae (Falls lukunani), 227–229, 228–229*f*
Cichla intermedia (royal peacock bass), 102–103, 102–103*f*
Cichla melaniae (Xingu peacock bass), 188–189, 189*f*
Cichla mirianae (Fire peacock bass), 212–214, 212–214*f*
Cichla ocellaris (butterfly peacock bass), 57–62
Cichla pinima (Pinima peacock bass), 150–152, 151–152*f*
Cichla piquiti (Blue peacock bass), 169–170, 169–170*f*
Cichla temensis (speckled peacock bass), 130–134, 131–133*f*
Reservoir stocking, 74
Retroclus xinguensis, 233*f*
Rewa River (Essequibo), 38*f*, 261*f*
Río Cinaruco, 1
Rio Negro, 34–35, 38*f*, 40, 44*f*, 46–47, 46*f*, 51, 58*f*, 72–73, 74*f*, 75–77, 76*f*, 79*f*, 84–85, 106–107, 110, 115, 118, 120*f*, 126, 127*f*, 251–254, 251–253*f*, 274*f*
Rocky shoals, 216, 217*f*, 219*f*, 222, 224*f*, 226–227, 228*f*
Royal peacock bass. *See Cichla intermedia* (royal peacock bass)
Rupununi River, 35–36, 36*f*, 51, 54, 56, 263*f*

S

Santos Luzardo National Park, 1, 269–270, 270*f*
São Benedito River, 194–195, 194–196*f*, 197, 198*f*
Semaprochilodus kneri (bocachicos), 269–270
Serrasalmus species (Piranhas), 7*f*, 107
Severums (*Heros severus*), 66–67, 138
Siapa River, 272*f*
Silver dollars (*Metynnis* spp.), 165
Small tetras (*Hemigrammus* spp. and *Hyphessobrycon* spp.), 165
Spatial food-web subsidy, 4
Species geographic distribution, 234–237, 239
Speckled peacock bass. *See Cichla temensis* (speckled peacock bass)
Spectacled caiman (*Caiman crocodilus*), 82, 82*f*
Sport fishing tourism, 273–274, 277, 280, 287–288
Stocking, 163–164
Subsistence fishery, 274, 276, 289–291, 293*f*

T

Tapajos River, 254–255, 254–255*f*
Tapirus terrestris (tapir), 197*f*
Taxonomy
Cichla, 34–35, 39, 138, 143
Cichla cataractae (Falls lukunani), 216
Tetras (Characidae), 7*f*
Threatened species, 175–176, 177*f*
Tiger shovelnose catfish (*Pseudoplatystoma tigrinum*), 138
Trophy fish, 110
Tucunaré fogo, 194–195
"Tucunaré", 8–9, 154–156

U

Upper Paraná River (Tocantins), 38*f*, 160*f*, 164
Upper Tapajos Basin, 194*f*, 202*f*, 208*f*

V

Vampire fish (Cynodontidae), 7*f*
Venezuela, 67, 67–68*f*, 70, 71*f*, 72–74, 83–84, 92–93, 92*f*, 97–98, 97–98*f*, 100*f*, 101, 102*f*, 220–221*f*, 222, 223*f*, 227. *See also Cichla orinocensis* (Orinoco butterfly peacock bass)

Vertical dark bars, 26–27, 27*f*
Ventuari River, 293*f*
Volta Grande (Big Bend) stretch, 175–176

X
Xingu peacock bass. *See Cichla melaniae* (Xingu peacock bass)
Xingu River, 174, 174–175*f*, 255–259, 256–259*f*

W
Water quality parameters, 14*t*
Wolf fish (Erythrinidae), 7*f*

Y
Yapacana National Park, 276*f*

Z
Zebra pleco (*Hypancistrus zebra*), 177*f*
Zoogeography, 240–263, 241–242*f*